Herausgeber: Markus Greim, Wolfgang Kusterle und Oliver Teubert

Rheologische Messungen an Baustoffen 2019

Tagungsband zum 28. Workshop und Kolloquium, 13. und 14. März an der OTH Regensburg

ISBN (Paperback): 978-3-7482-4651-0

ISBN (Hardcover) :978-3-7482-4652-7

ISBN (e-Book) : 978-3-7482-4653-4

1. Auflage 2019

Layout: Benedikt Hoffmann

Satz: Christian Greim, Dr. Helena Keller

Verlag und Druck: tredition GmbH, Halenreie 40-44, 22359 Hamburg; www.tredition.de

Publisher: Markus Greim, Wolgang Kusterle and Oliver Teubert

Rheological Measurement of Building Materials 2019

Proceedings of the 28th Conferences and Laboratory Workshops, 13th and 14th March at OTH Regensburg

ISBN (Paperback): 978-3-7482-4651-0

ISBN (Hardcover) :978-3-7482-4652-7

ISBN (e-Book) : 978-3-7482-4653-4

1. Edition 2019

Layout: Benedikt Hoffmann

Typesetting: Christian Greim, Dr. Helena Keller

Published and printed by: tredition GmbH, Halenreie 40-44, 22359 Hamburg; www.tredition.de

Preface

Rheological measurements help to optimize the robust application of building materials. For the past few years applied rheology has been moving increasingly into the center of attention of the civil engineering society. Advanced testing equipment, intensive research and development work have all helped to deepen the understanding of this field. Last but not least, different conferences, such as the 28th Conference on the Rheology of Building Materials at the OTH, have helped to exchange the findings and establish valuable new contacts. Finding a compromise between rheological theory and its practical application has been the main topic of these 28 conferences.

In these proceedings important rheological topics are addressed: specific applications of concrete such as self-compacting concrete, tremie concrete and sprayed concrete and mortar mixes for 3D printing, which all may be classified as self-compacting concretes. The consideration of time- and temperature-dependent changes in workability are essential for a successful application of these materials. Furthermore, the influence of the mixing energy used and the way of casting/spraying is of high importance for the rheological properties. The rheology of other concrete types is drastically changed due to the compaction by vibrators.

Concrete rheology is controlled by its paste content and the rheology of the paste. Tests on pastes are easier to perform, as particle migration of aggregates is not present. However, one has to ask whether it is sufficient to use the results from the paste-measurements for the assessment of concrete and how the optimized paste content will be predetermined. The other question is which admixtures or additives may help to improve the very different application processes. Apart from using superplasticizer and stabilizers, entrained air is a powerful tool to influence rheology. One has to consider, though, how entrained air may behave under different levels of pressure in the pumped concrete.

Beside concrete, other papers deal with highly flowable materials such as injection grout as well as grouting mortar and gypsum based adhesive mortars.

I would like to thank all authors and conference delegates for their contribution as well as all members of the organizing committee for their support. This event is organized as a low-cost conference without any attendance fee. This is only possible due to numerous hours of work put in by the organizing committee, Schleibinger Testing Systems and the hospitality of the OTH.

We are looking forward to meeting you in Regensburg again.

Wolfgang Kusterle

Scherinduzierte Partikelmigration: Relevanz bei rheologischen Messungen an Mörteln und Betonen

Christian Baumert, Prof. Harald Garrecht
Institut für Werkstoffe im Bauwesen, Stuttgart
Tel.: 0711-68562794, E-mail: Christian.Baumert@iwb.uni-stuttgart.de

Kurzfassung

Um die rheologischen Eigenschaften von Betonen und Mörteln messen zu können, müssen zahlreiche Randbedingungen eingehalten werden. So gilt es unter anderem die Homogenität der Probe während der Messung sicher zu stellen. Die Scherbelastung führt zu einer Migration der groben Partikel in Bereiche niedriger Scherbelastung, und somit zu einer Entmischung. Um jedoch den Strukturabbau von zementösen Baustoffen vor der eigentlichen Messung zu erreichen, ist eine Vor-Scherphase unabdingbar. Der dabei unvermeidlichen Entmischung der Probe konnte bisher nur durch Messwerkzeugen begegnet werden, die nicht den Anforderungen an Messwerkzeugen zur Messung in absoluten Einheiten entsprechen. Eine Umrechnung der Rohdaten in absoluten Einheiten, mit diesen sogenannten relativen Messwerkzeugen, ist nicht statthaft. Werden sogenannte Absolut-Messwerkzeuge eingesetzt, erfolgt die Messung nach der Vorscherung jedoch in einer nicht homogenen Probe. In diesem Beitrag wird ein Messsystem vorgestellt, dass die Vorscherung der Probe mit aktiver Homogenisierung kombiniert und somit die Messungen mit Absolut-Messwerkzeugen an homogenen Proben ermöglicht. Vergleichende Messungen zwischen dem neuen Messkonzept und dem klassischen Messregime belegen erhebliche Abweichungen bei den gemessenen Rohdaten. Folglich wird die scherinduzierte Partikelmigration aktuell bei Messungen an granularen Suspensionen mit Absolut-Messwerkzeugen nicht ausreichend berücksichtigt und führt zur massiven Unterschätzung von Fließgrenze und Viskosität in absoluten Einheiten.

1 Einleitung

Bei vielen modernen Betonen ist die Messung der rheologischen Eigenschaften unmittelbar nach Mischende nicht ausreichend, da sich diese bis zum Einbau häufig verändern. Aufgrund des thixotropen Verhaltens bildet sich eine Struktur aus, welche die Fließgrenze und die Viskosität ansteigen lassen. Nach Aufbringung einer Scherbelastung wird diese Struktur – in Abhängigkeit von der Höhe der Scherbelastung – teilweise oder vollständig wieder abgebaut.

Um das rheologische Verhalten der Mörtel und Betone während der Mischungsentwicklung unter Praxisbedingungen vorhersagen zu können, wären folglich mehrere Messungen zu unterschiedlichen Zeitpunkten bei variabler Vorscherungsintensität wünschenswert. Der Setzfließmaß-Versuch bzw. der VdZ-Tricher sind hierbei als unzureichend einzustufen. Diese achsen-symmetrischen 1-Punkt-Versuche ermöglichen aufgrund des geringen Probenvolumens nicht die für eine zielgerichtete Entwicklung nötige Bestimmung der Fließgrenze in absoluten Einheiten. Roussel [1] hat ein nicht praxistaugliches erforderliches Probevolumen für diese Versuche von 300l berechnet.

Beton-Rheometer mit Absolut-Messwerkzeugen ermöglichen wiederholte Messungen bei veränderlichen Scherbelastungen und Vorscherung. Allerdings haben mehrere Rundversuche mit unterschiedlichen Beton-Rheometern erhebliche Abweichungen bei den berechneten rheologischen Parametern offengelegt. Nach [2] müssen diese massiven Abweichungen auf die Verletzung einer oder mehrerer Haupthypothesen für die Umrechnung der Maschinendaten Drehmoment und Drehzahl in die lokalen Daten Scherspannung und Scherrate beruhen.

- Wandhaftung des Probematerials an beiden Oberflächen im Scherspalt
- Laminares Fließgleichgewicht im Scherspalt
- Homogenität des Probematerials

Durch eine raue bzw. leicht profilierte Oberfläche lässt sich das Gleiten der Probe an den Messinstrumenten wirkungsvoll unterbinden. Die Verfälschung der Messwerte durch eine leichte Profilierung der Oberfläche ist gegeben, ist größenordnungsmäßig jedoch sicherlich unerheblich, und kann als Begründung, für die Abweichungen bei den Rundversuchen, ausgeschlossen werden. Bezüglich Fließ- und Materialhomogenität bestehen allerdings erhebliche Unsicherheiten. Mörtel und Betone verfügen immer über eine Fließgrenze, insofern muss der kritische Radius (Übergang nicht Fließen/Fließen) bei den Berechnungen immer berücksichtigt werden. In der Literatur wurde das Phänomen der scherinduzierten Partikelmigration bei feststoffreichen Suspensionen lange als sehr langsamer Vorgang bewertet und somit als irrelevant bei rheologischen Messungen erachtet. [2] benennt eine alte kritische Scherrate, die sich bestimmt aus Scherspaltweite^2/Partikelgröße^2. Bei 16 mm Größtkorn und einer 4-fachen Scherspaltweite von 64 mm ergibt sich eine kritische Scherrate von 16 s^{-1}. Eine solch hohe

Scherrate wird in Beton-Rheometern nicht verwendet. Allerdings haben [3] aufgezeigt, dass - bei hoher Packungsdichte und in Abhängigkeit von den Fließeigenschaften - bereits bei einem Fünfhundertstel der berechneten Scherrate sehr schnelle partikelinduzierte Migration erfolgt und somit unvermeidbar ist. Für einen Mörtel haben [4] nach einer sehr schnellen Aufwärtsrampe bedeutsame Unterschiede in der radialen Verteilung des Sandes bestimmt. Folglich ist die scherinduzierte Partikelmigration in der Vorscher-Phase unvermeidlich und sind somit bei der Auswertung von rheologischen Messungen an Mörteln und Betonen zwingend zu berücksichtigen.

2 Lösungsansatz

Um die Entwicklung von Mörteln und Betonen praxisnah und schnell umsetzen zu können, wurde der Labormischer KNIELE KKM-RT 22.5/15 entwickelt. Erstmalig ist es möglich, Mörtel und Betone mit stufenlos wählbarer Mischintensität in einem sogenannten Intensivmischer herzustellen und im Nachgang rheologische Messungen im Mischer durchzuführen. Unabhängig vom Hersteller zeichnen sich Intensivmischer dadurch aus, dass in einem Teilvolumen des Mischgutes mit sehr hoher Werkzeuggeschwindigkeit und somit einhergehender sehr hoher Maschinen-Froudezahl Partikelagglomerate aufgelöst werden können. Dadurch kann die Fließfähigkeit des Materials – insbesondere die Viskosität – gegenüber Standardmischern positiv beeinflusst werden. Nach dem eigentlichen Mischvorgang kann der Nutzer entscheiden, ob die rheologischen Messungen mit dem Mischwerkzeug oder aber einem Messwerkzeug durchgeführt werden sollen. Die Messungen mit dem Mischwerkzeug können ohne Werkzeugwechsel durchgeführt werden. Da die Oberflächen des Mischwerkzeugs auf die Mischgutbewegung ausgelegt sind, ist eine Festlegung der Werkzeugoberfläche und des beanspruchten Probevolumens nicht möglich. Die Messwerte Drehzahl und Drehmoment können nicht in absolute Einheiten umgerechnet werden, besitzen also relativen Charakter. Diese relativen Messwerte sind für eine Kontrolle der Gleichmäßigkeit der einzelnen Chargen in der Praxis bei Produktionsmischern ausreichend. Soll jedoch eine gezielte Mischungsentwicklung erfolgen, sind Messwerte in absoluten Einheiten erforderlich. Dazu können die Mischwerkzeuge des KKM-RT 22.5/15 nach dem Mischvorgang durch Schnellwechselvorrichtungen gegen Messwerkzeuge ausgetauscht werden, deren Oberflächen definiert sind. Der innere Antrieb, eine High-Torque Synchronmotor, treibt dabei den Messzylinder an. Die Drehzahl wird über einen hochauflösenden Drehgeber erfasst, das Drehmoment über einen Frequenzumrichter, der nach dem Direct-Torque-Prinzip arbeitet. Der äußere Zylinder wird durch den Antrieb des Abstreifers angetrieben und wird während der eigentlichen Messung durch eine Motorbremse in seiner Position fixiert. Das Novum besteht in der Spirale, welche auf der Außenseite des äußeren Zylinders angebracht ist. Damit kann das Material mit dem äußeren Antrieb zwischen Außenzylinder und Innenwandung des konusförmigen Mischbehälters nach oben transportiert werden. Am oberen Ende der Spirale fällt das Material durch Öffnungen in den Scherspalt zwischen innerem und äußerem Zylinder.

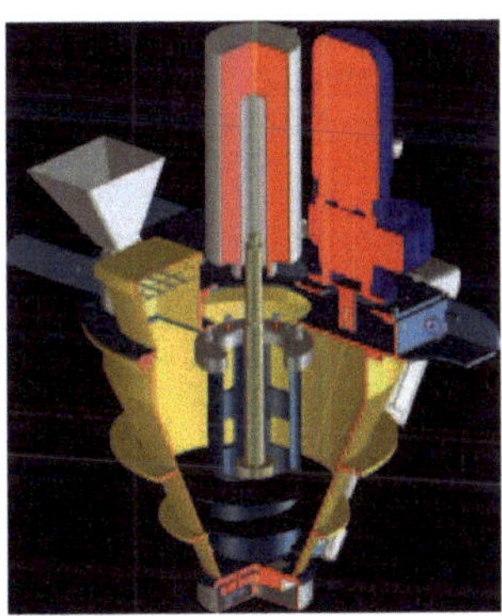

Abbildung 1: Absolut-Messwerkzeuge des KNIELE KKM-RT 22.5/15

Ergebnis ist die 3-dimensionaler Umlauf und die Vorscherung des Materials. Die Vorscherung selbst findet also nicht über den inneren Zylinder statt, sondern über die Spirale. Einer scherinduzierten Partikelmigration innerhalb des Scherspaltes wird also aktiv entgegengewirkt und eine der Haupthypothesen bei der Umrechnung auf absolute Einheiten nachweislich erfüllt. Über den in der Bodenklappe integrierten Feuchte-Sensor wird während der Umwälzung des Materials mit der Spirale die Feuchtigkeit des Materials in einem Teilvolumen unmittelbar vor der Feuchte-Sonde bestimmt. Da die Messwerte während des Umwälzvorgangs nur minimal schwanken, ist es statthaft, die gleichmäßige Feuchteverteilung im Mischgut und somit die Homogenität, anzunehmen.

3 Versuchsdurchführung

Um den Einfluss der scherinduzierten Partikelmigration aufzuzeigen, wurde ein leichtverdichtbarer Beton nach dem sogenannten Smart Dynamic-Konzept der BASF entwickelt und zwei unterschiedlichen Messregimen unterzogen.

Tabelle 1: Zusammensetzung des untersuchten leicht verdichtbaren Betons

	[kg/m³]
CEM III/A 42.5 N	292
Kalksteinmehl s&h easyflow	98
Fließmittel BASF ACE 430	2,2
Stabilisierer BASF Rheomatrix	0,88
Wasser	175
Rheinkies 8/16	381,5
Rheinkies 2/8	606,3
Rheinsand 0/2	783,8

Beim ersten Messregime erfolgt die Vorscherung (Dauer 5 s) bei der maximalen Scherrate des anschließenden Versuches über den inneren Messzylinder. Beim eigentlichen Messvorgang wird die Scherrate stufenweise (nach jeweils 5 s) abgesenkt. Dieses Vorgehen entspricht dem klassischen Messregime eines Beton-Rheometers. Beim zweiten Messregime erfolgt die Vorscherung über die Spirale für 30 Sekunden vor dem Messvorgang und vor jeder Drehzahländerung. Somit liegt bei diesem Messregime vor jeder Drehzahländerung ein homogenisiertes Material vor.

3.1 Auswahl der Messdaten

Nach der Drehzahländerung muss sich zunächst ein Fließgleichgewicht im Scherspalt ausbilden. Dieser Vorgang wird jedoch durch die scherinduzierte Partikelmigration überlagert. Deshalb wurden die Mittelwerte für das Drehmoment 0.5 s – 1.5 s nach Drehzahlwechsel, 1 s – 2 s nach Drehzahlwechsel und 2 s – 3 s nach Drehzahlwechsel bestimmt.

- $SV_{0.5\text{-}1.0}$: Vorscherung mit der Spirale vor der Messung und vor jedem Drehzahlwechsel über 30 Sekunden; Mittelwert aus den Drehmoment-Messwerten 0.5 bis 1.0 Sekunden nach Messbeginn.

- $SV_{1.0\text{-}2.0}$: Wie zuvor, aber Mittelwert aus den Drehmoment-Messwerten 1.0 bis 2.0 Sekunden nach Messbeginn.

- $SV_{2.0\text{-}3.0}$: Wie zuvor, aber Mittelwert aus den Drehmoment-Messwerten 2.0 bis 3.0 Sekunden nach Messbeginn.

- $RM_{0.5\text{-}1.0}$: Vorscherung mit dem inneren Messwerkzeug vor dem Messvorgang. Mittelwertbildung aus den Drehmoment-Messwerten 0.5-1.0 Sekunden nach Messbeginn.

- $RM_{1.0\text{-}2.0}$: Wie zuvor, aber Mittelwertbildung aus den Drehmoment-Messwerten 1.0-2.0 Sekunden nach Messbeginn.
- $RM_{2.0\text{-}3.0}$: Wie zuvor, aber Mittelwertbildung aus den Drehmoment-Messwerten 2.0-3.0 Sekunden nach Messbeginn.

Um das Leerlaufdrehmoment des Antriebs und den Drehmomentanteil zu bestimmen, welcher durch die Bodenfläche des Messzylinders im Beton verursacht wird, wurden weitere Messungen durchgeführt. Dazu wurde der Antriebsteil des Mischers so weit nach oben verfahren, bis nur noch die Bodenfläche des Messzylinders Kontakt zum Beton hatte. Dann wurden die einzelnen Drehzahlstufen abgefahren und Mittelwerte für das benötigte Drehmoment gebildet. Die ermittelten Werte wurden bei den rheologischen Messungen in Abzug gebracht und ermöglichen somit den Bezug nur auf die Mantelfläche des Messzylinders.

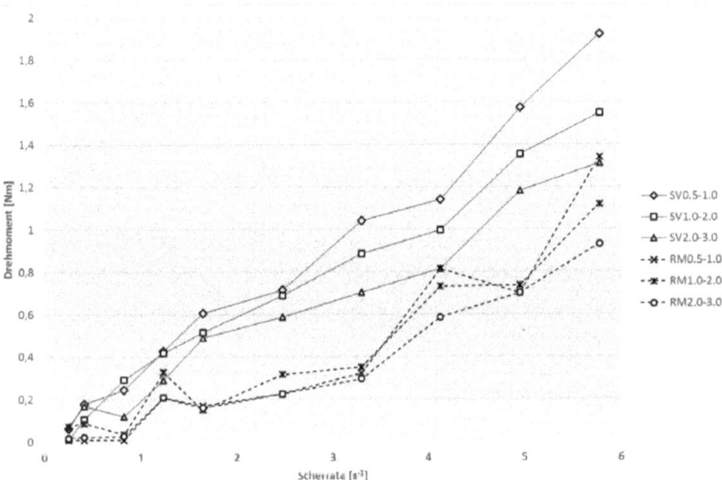

Abbildung 2: Darstellung der Drehmoment-Mittelwerte (Netto-Werte) nach Abzug der Leerlaufverluste für die beiden unterschiedlichen Messregime

Nach [1] ist gegenwärtig nicht bestimmbar ist, welches Beton-Rheometer die richtigen absoluten Werte ausgibt. Da bei der LCPC-Box nach gegenwärtigem Kenntnisstand alle Randbedingungen zur Bestimmung der Fließgrenze in absoluten Einheiten erfüllt sind, wurde diese als Referenz festgelegt. Bei einer Fließstrecke von L = 62 cm ergibt sich eine Fließgrenze von ca. 70 Pa.

3.2 Berechnung der lokalen Daten

Um die Maschinendaten Drehzahl und Drehmoment in absolute Einheiten zu überführen, wurde die Methode von Reiner Riwlin gewählt. Als Eingangsdaten wurden für beide Messregime die Mittelwerte der Drehmomente in den oben genannten drei Intervallen gewählt (in Abbildung 2 dargestellt).

Tabelle 2: Gegenüberstellung der nach Reiner Riwlin berechneten rheologischen Bingham-Parameter für die beiden Messregime

	SV0.5-1.0	SV1.0-2.0	SV2.0-3.0	RM0.5-1.0	RM1.0-2.0	RM2.0-3.0
Fließgrenze in [Pa]	0,010	23,141	4,140	0,010	0,010	0,010
Viskosität in [Pas]	65,368	51,767	46,309	39,155	35,202	29,485
Abweichung Viskosität (Basis $RM_{1.0\text{-}2.0}$)	+ 85 %	+ 47%	+ 31 %	+ 11%	0 %	- 16 %
Bestimmtheitsmaß R^2	0,992	0,995	0,987	0,933	0,947	0,969
min. Geschwindigkeit ohne Scherlokalisation [rad/s]	0,000	0,542	0,108	0,000	0,000	0,000

Das traditionelle Messregime (Vorscherung vor der eigentlichen Messung durch das Messwerkzeug) führt zu den geringsten berechneten Viskositäten. Wird die Probe vor jeder Drehzahländerung durch die Spirale auf dem Außenzylinder homogenisiert, steigt die berechnete Viskosität an. Mit der LCPC-Box wurde eine Fließgrenze von ca. 70 Pa bestimmt. Die in der obigen Tabelle ausgewiesenen Lösungen mit einer Fließgrenze von nahezu Null sind also zu verwerfen. Das klassische Messregime mit der Auswertung nach Reiner Riwlin ist folglich nicht geeignet den untersuchten Beton hinsichtlich Frischbetoneigenschaften zu beschreiben. Die Auswertung von $SV_{1.0\text{-}2.0}$ liefert lediglich eine Fließgrenze von 23 Pa und weicht somit noch immer erheblich von der LCPC-Box ab.

3.3 Interpretation und Ausblick

Die Ergebnisse in der Tabelle 2 lassen sich durch die Literatur [2, 3 und 4] erklären. Infolge der sehr schnell ablaufenden scherinduzierten Partikelmigration erfolgt die Bestimmung der Messwerte an nicht homogenen Proben. Fließgrenze und Viskosität werden erheblich unterschätz. [4] sieht in diesem Vorgang die wahrscheinlich größte Fehlerquelle in der Bestimmung von rheologischen Eigenschaften an zementösen Proben. Mit der Messeinrichtung des KNIELE KKM-RT 22.5/15 kann dem erstmalig wirkungsvoll begegnet werden. Zwar liegt die berechnete Fließgrenze für den untersuchten Beton auch hier unter dem Referenzwert von ca. 70 Pa, doch ist dies wahrscheinlich auch auf den bei Beton-Rheometern üblichen drehzahlgeführten Betrieb zurückzuführen. Dabei wird die Fließgrenze über das rheologische Modelle extrapoliert. Nach [5] „sollte diese Methode nur für einfache QS-Tests verwendet werden, aber nicht mehr für zeitgemäße Forschung und Entwicklung". Beim KKM-RT 22.5/15 wird deshalb zukünftig im Forschungsvorhaben die Schubspannung vorgegeben und die Fließgrenze somit direkt als Messwert bestimmt.

4 Danksagung

Besonderer Dank gilt der DFG für die Förderung des Institutes für Werkstoffe im Bauwesen der Universität Stuttgart im Rahmen des Schwerpunktprogrammes 2005.

Literatur

[1] Roussel, N.: From industrial testing to rheolgical parameters for concrete, Understanding the rheology of concrete, Seiten 83-95, 2012, Woodhead Publishing

[2] Ovarlez, G.: Introduction to the rheometry of complex suspensions, Understanding the rheology of concrete, Seiten 23-62, 2012, Woodhead Publishing

[3] Fall, A. et al.: Shear thickening and migration in granular suspensions, PhysRevLett, 105, 268303

[4] Hafid, H. et al.: Estimating measurement artifacts in concrete rheometers from MRI measurement on model materials, RILEM Book Series, 2010, 1, Seiten 127-137

[5] Mezger, T.: Das Rheologie Handbuch, 4. Auflage, 2012, Vincentz Network.

Effect of the mixing time on the rheological parameters of cement pastes

Mareike Thiedeitz, Thomas Kränkel, Christoph Gehlen
Technical University of Munich, Centre for Building Materials (cbm), Munich, Germany; Phone: +49-8928927119, e-mail: mareike.thiedeitz@tum.de

Abstract

Cement pastes are colloidal suspensions whose flow properties are dependent on the presence of attractive and repulsive forces between the particles and due to that, on the bias of the colloidal system to flocculation and coagulation. Mixing time and intensity and thus the applied shear to the system changes the interparticle network in the cement paste which affects the rheological properties yield stress and viscosity significantly. Due to that, different preshear times of 30 to 300 s before the measurements were tested for cement pastes and the stress-strain curves and thus the rheological parameters determined. It could be observed that the yield stress and viscosity of the pastes decrease with increasing mixing times due to a higher de-flocculated state at the beginning of the rheological measurements.

1 Introduction

Rheology of cementitious pastes is a matter of interest in concrete technology since many years. In suspension rheology, cement paste is a two-phase suspension with water as liquid and cement particles as solid phase. The size of cement particles range from app. 10^{-3} to 10^{2} µm, thus a cement paste is a colloidal suspension which exhibits a yield stress and behaves as a Non-Newtonian liquid with changing plastic viscosity under applied shear due to changing interparticle forces. The rheological parameters yield stress, viscosity and thixotropy of the cementitious suspension affect the flow behavior of concrete which in turn determines the slump, flow and therefore workability in the field [1–3]. Thus the knowledge about these parameters is of special interest for the determination of the flow behavior of Self Compacting Concrete (SCC) and Ultra-High Performance Concrete (UHPC) which contain high amount of colloidal and fine particles. Due to colloidal interactions and chemical nucleation hydration effects of these fine particles, the prediction of flowability of these concretes gets more difficult.

Generally, viscosity and yield stress increase with decreasing particle diameters and increasing solid contents in the suspension. Cement particles form microstructures which affect the effective apparent granulometry: In a completely deflocculated suspension, particles are separated whereas with ongoing formation of interparticle networks, effective particle diameters increase [4, 5]. Therefore, with a change of microstructure the apparent viscosity of the suspensions changes [4–7].

A change of microstructure can be induced by shear. The rate and time of shear meanwhile affects the grade of microstructural change and thus the apparent viscosity which has to be taken into account in rheological measurements. For correct rheological measurements, a reference state of flocculation of the tested paste has to be found. Two possible ideal reference states would be either a completely flocculated or a completely deflocculated system. To correctly determine the formation of interparticle networks due to Brownian motion and colloidal interactions, a completely de-flocculated suspension is the most appropriate reference state. However, a completely deflocculated state will never be obtainable due to the strong formation of particle networks in a few seconds directly after mixing [7]. Therefore, the most reasonable reference state has to be found which thereupon can be called the most de-flocculated state [8]. Imaging cement pastes after an appropriate time of rest, the structure can be called a virgin structure of a completely flocculated network. Applied shear results in a structural breakdown until equilibrium is reached where the applied shear is not powerful enough to break the flocs. The time until this steady state is reached clearly depends on the grade of applied shear. If a high shear is applied, more de-flocculation can take place and the microstructure of the suspension will end up in smaller flocs. Still, in cement pastes it has to be taken into account that the applied shear will cause additional energy in the system due to friction between the particles which will change the chemical reaction behavior of the cement. According to that, in cementitious colloidal suspensions, the full structural breakdown of an already built reversible network should never be attended. An appropriate compromise of floc breakage due to the preshear before a rheological measurement should thus be found for experimental procedures. Still, the grade of remaining particle network in the microstructural system has to be known for a precise formulation of possible structural build up and breakdown.

2 Rheological background in suspension rheology

Colloidal suspensions generally are two-phase fluids which contain a liquid and a solid phase with a particle diameter from 10^{-3} to 10^{2} µm [7]. Possible interaction forces in microscopic suspensions are hydrodynamic forces due to relative motion, Brownian motion of particles smaller than 10 µm due to thermal randomizing activity and colloidal interaction forces which are potential forces and can be either attractive or repulse. Attractive forces namely are Van-der-Waals-forces, electrostatic attraction, attractive hydrophobic and bridging forces whereas repulsive forces are caused by electrostatic repulsion and steric hindrances [7–12]. The main apparent forces between colloidal particles originate from their physical and chemical properties. Especially hydrodynamic forces originate from particle motion and increase with rising particle diameters. Especially in suspensions with inert particles, hydrodynamic forces affect the rheophysical flow behavior of the suspension. Depending on the sum of contact forces which are caused by collisions and non-contact colloidal forces, particle interactions either lead to stable dispersions in the fluid due to predominant repulsion forces or to the formation of agglomeration networks due to flocculation or coagulation [4, 5, 9, 11, 13, 14].

With increasing fineness of reactive powders, interparticle forces tend to increase. Thus, the change of particle size of solids can change the rheological parameters tremendously while remaining at the same solid concentration [3, 7]. Cement pastes are colloidal suspensions with particles from a few nm to app. 100µm [8, 15]. Due to the colloidal attractive forces, cement particles flocculate to strong interparticle networks [3, 16]. The presence and magnitude of particle networks due to interparticle forces determines the rheological parameters yield stress and viscosity as well as the bias for reversible network breakage and build up, called thixotropy. Shear-induced changes in microstructure lead to either shear-thinning or shear-thickening behavior.

2.1 Origin of yield stress

The yield stress of cement pastes originates from chemical and physical interactions between the particles which tend to be a particle network due to contact and colloidal forces [9, 17, 18]. In cementitious suspensions, interparticle non-contact surface forces predominate over mechanical forces and tend to particle agglomeration [19]. Due to these interparticle forces, particle agglomerates retain elastic stress in the particle network which has to be overcome to make the suspension flow. Below this value, the flocs in the particle network retain an elastic deformation without flow [5]. A completely deflocculated suspension and therefore no elastic response would theoretically result in a viscous Newtonian fluid without a yield stress [20]. Contrary, any formed microstructure in particle suspensions will produce a yield stress. The predominant interparticle forces in a cementitious suspension are dependent on different parameters. The particle size distribution and the relative solid concentration Φ_{rel} (ratio of the solid concentration to the maximum solid concentration Φ/Φ_{max}) for example determine the distance of the particles and thus the value of Van-der-Waals forces [15]. Superplasticizers cause enhanced particle dispersion due to repulsive electrostatic forces and steric hindrance and thus lead to lower yield stress [21–23].

2.2 Origin of viscosity

The viscosity of a suspension is the resistance against shear. If the viscosity is constant, the fluid is called Newtonian fluid. Due to particle interactions, the viscosity can change with increasing shear. Consequently the liquid is Non-Newtonian and can either behave shear-thinning or shear-thickening. Viscosity is mainly determined by the relative solid concentration Φ_{rel} and particle size distribution. With increasing relative solid concentration Φ_{rel}, viscosity increases due to decreasing interparticle distances and thus increasing colloidal interactions and contact forces respectively [24]. Still, it is possible to reduce the shear viscosity tremendously with a change in the particle size distribution while keeping the relative solid concentration constant. With decreasing average particle diameters, viscosity increases due to a higher particle surface area and therefore a bigger value of interparticle forces which gain a bigger potential for the formation of particle agglomerates [14, 25, 26].

2.3 Effect of shear history on rheology

In structuration kinetics researches, it was already investigated that flocculation and coagulation kinetics strongly depend on the initial shear, the shearing time and the energy of applied shear [8, 9, 20, 27]. In an aggregated suspension, shear works as a dispersion force which is able to break the flocs or aggregates. With increasing shear rate, the floc size decreases. As long as the applied shear is strong enough to break the flocs, viscosity will decrease until equilibrium is obtained. The viscosity thus will decrease asymptotically at high shear rates (see Figure 1). Meanwhile, if the shear is limited and not strong enough to break the flocs, suspensions will remain with bigger flocs or clusters due to the strength of agglomeration [26, 28]. Moreover, shear-induced particle collisions produce re-formation of networks [20]. After shear, particles will agglomerate again. The strength of the rebuild networks and the rate of structural build up then depend on the agglomeration state and the apparent particle size distribution. Thus, the particle network is a function of the mixing history of the cement paste.

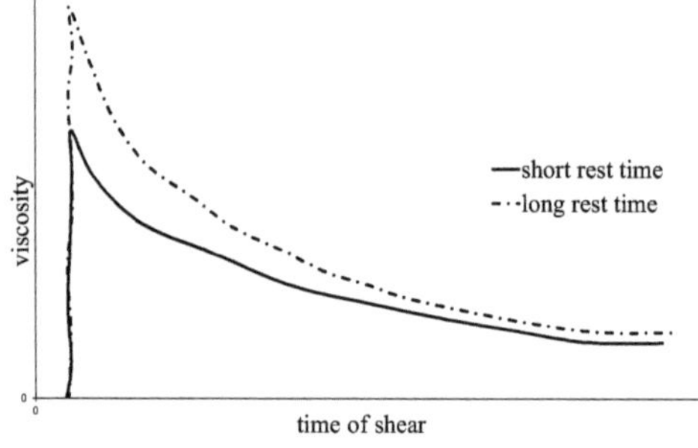

Figure 1: Viscosity as a function of shearing and resting time before shear acc. to [26]

3 Materials and Methods

3.1 Materials and mix design

Cement pastes were mixed using an Ordinary Portland Cement CEM I 42.5 R according to EN 197-1 and demineralized water with an adjusted temperature for a fresh cement paste temperature of 20 °C. Cement pastes with w/c ratios of 0.40 and 0.42 and thus solid concentration 0.434 and 0.446 were produced for a targeted variation of the paste viscosity For the variation of the yield stress at the same relative solid concentration a PCE-type superplasticizer was used. The PCE has a water content of 77 % which is subtracted from the water addition in mixture 3. The paste mixtures can be seen in Table 1.

Table 1: Cement paste mixtures

Mixture	w/c ratio	Solid concentration [-]	Cement [kg/m³]	Water [kg/m³]	PCE [wt.-% by cement]
0.42	0.42	0.434	1346.7	565.6	-
0.40	0.40	0.446	1383.9	553.6	-
0.40-SP	0.40	0.446	1383.9	551.3	0.25

The mixing of the pastes was done by a mortar mixer using a mix regime according to EN 196-1. The preshear before the rheological measurements was performed using a standard drilling machine with a propelling screw. The rheological experiments were performed using an Anton Paar Rheometer MCR 502. Rheological measurements were conducted using a parallel plate device with a plate diameter of 50 mm and a double serrated surface to prevent wall slip. The gap during measurements was set to 1 mm to be enable to quantify absolute values for the yield stress and viscosity respectively.

3.2 Experimental procedure

The time of water addition was set as starting time. The mixing duration was 4 min in total and started without delay after water addition. After mixing, the cement paste was left at rest for some time before preshear. For every mixture, single batches were tested with preshear times of 30, 90, 150 and 300 s. The preshear time with a duration of 30 s started 12:30 min after water addition. Longer preshear times thus started earlier so each preshear time ended 13:00 min after water addition. Cement paste then was placed in the rheometer. The shear rate profile is shown in Figure 2. Again, a 30 s preshear with a constant shear rate of $\dot{\gamma} = 40\ s^{-1}$ was passed to break the structure caused by resting time between sample placement in the rheometer and the start of the rheometric measurement. Following, a shear rate profile with stepwise decreasing shear rates from $\dot{\gamma} = 80\ s^{-1}$ to $\dot{\gamma} = 0.02\ s^{-1}$ was passed with a duration of 6 s for each step. Slump flow measurements using a mini slump cone acc. to EN 1015-3 were passed directly after starting the rheometric measurement. Each mixture was repeated twice for every preshear time. The results shown in the following sections are the average results of these two single measurements.

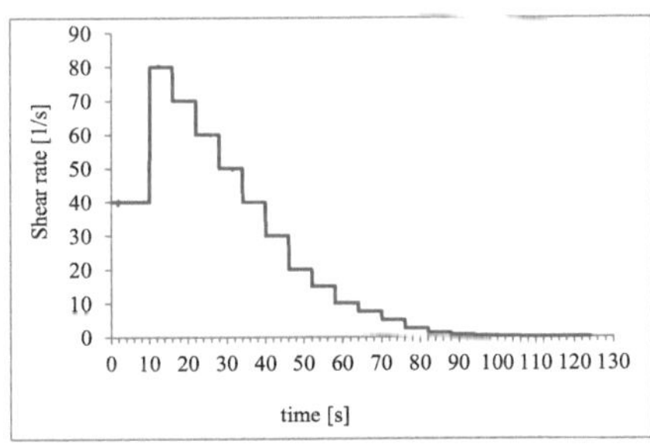

Figure 2: Shear rate profile for the rheological experiments

4 Experimental results

Figure 3 a shows the measured slump flow values and Figure 3 b - d the flow curves for the cement pastes 0.42, 4.40 and 0.40-SP as a function of the preshear time. The Herschel-Bulkley regression was used to calculate the yield stresses τ_0 and viscosities $\eta_{H.-B.}$ of the pastes based on the measured flow curves. The paste viscosities were exemplary taken from the Herschel-Bulkley regression at a shear rate of $\dot{\gamma} = 10\ s^{-1}$. The slump flow values SF as well as the calculated yield stresses τ_0 and viscosities $\eta_{H.-B.}$ of the cement pastes are given in Table 2.

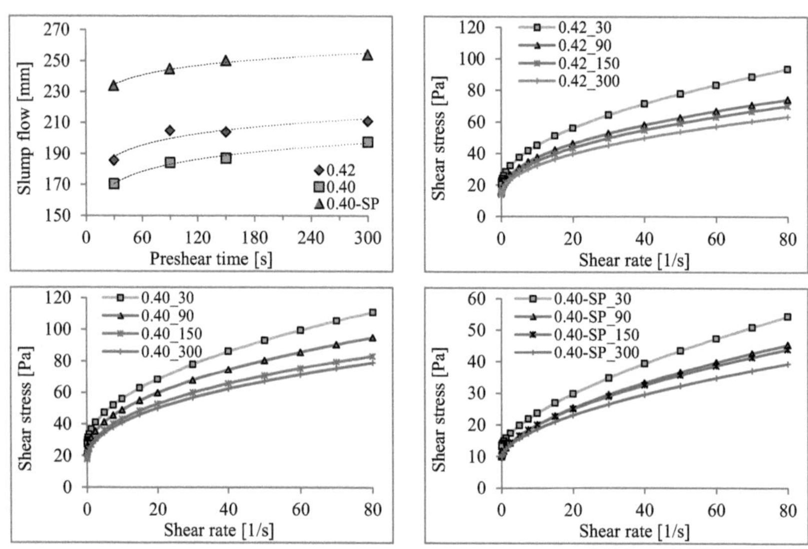

Figure 3: Slump flow values (a) and flow curves for the cement pastes 0.42 (b), 0.40 (c) and 0.42-SP (d) as a function of the preshear time

Table 2: Slump flow values and calculated yield stresses and viscosities for the cement pastes 0.42, 0.40 and 0.42-SP

Preshear time [s]	0.42			0.40			0.40-SP		
	SF [mm]	τ_0 [Pa]	$\eta_{H.-B.}$ [Pas]	SF [mm]	τ_0 [Pa]	$\eta_{H.-B.}$ [Pas]	SF [mm]	τ_0 [Pa]	$\eta_{H.-B.}$ [Pas]
30	185.6	19.61	1.30	170.5	25.96	1.50	234.0	13.08	0.69
90	204.8	13.65	1.07	184.0	20.4	1.32	244.9	9.77	0.61
150	204.0	12.63	1.02	187.0	16.03	1.18	250.2	9.84	0.59
300	211.0	12.18	0.90	197.6	15.0	1.12	254.0	9.56	0.52

In accordance with [29], the yield stresses of the pastes decrease with increasing slump flow values. However the decrease of the yield stress τ_0 for increasing preshear times is more pronounced than the increase in slump flow SF: The percentage decrease of yield stress is app. 33% for the pastes 0.42 and 0.42-SP and 40% for the paste 0.40 whereas the increase in slump flow varies from 9 to 16%: For example, for paste 0.42 the slump flow increased from 185.6 mm to 211.0 mm which is a percentage increase of app. 14%. For every paste mixture, the decline in yield stress is most pronounced within a preshear time of 90 s, prolonged preshear times don't lead to further striking decrease in yield stress. This is valid for every tested mixture.

The effect of the preshear time on the paste viscosity $\eta_{H.-B.}$ can be seen in Figure 4 and Table 2 respectively. With increasing preshear times, the viscosity decreases for every mixture. The most striking decrease takes place after 90 s. The difference in rheological parameters between 150 and 300 s gets little so it can be assumed that equilibrium of rheological parameters can be approached with longer preshear times. The regression curve for the viscosity decrease in relation to the preshear time follows a logarithmic function and is shown with dashed lines in Figure 4. It has to be taken into account that viscosity decreases with the addition of superplasticizer because of dispersed particles due to repulsive interparticle forces. The most pronounced decrease of viscosity takes again place within a preshear time of 90 s.

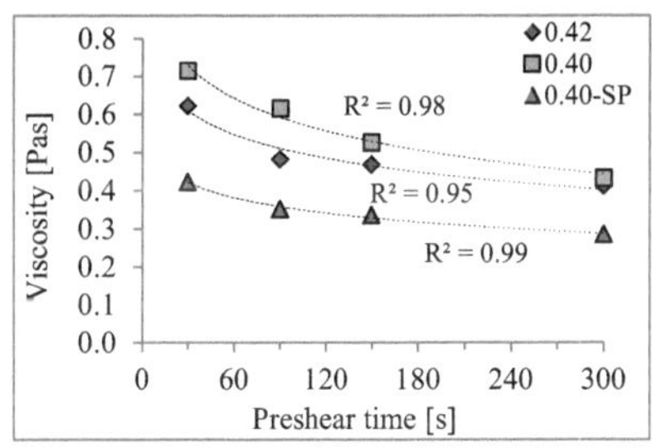

Figure 4: Viscosities of the cement pastes 0.42, 0.40 and 0.42-SP as a function of the preshear time

4.3 Discussion

It could be observed that the yield stresses τ_0 and the viscosities $\eta_{H.-B.}$ of the investigated pastes decreased with increasing preshear times because of the increased breakage of the particle network (cf. Figure 1). This fits the theory of the origin of a yield stress: With a stronger interparticle network, the yield stress increases due to higher elastic storage energy in the system which has to be surpassed to make the suspension flow.

The decrease in viscosity with increasing preshear time originates from a greater de-flocculation of the cement particle network. With increasing shear time, more flocs can be broken and an appropriate reference state can be reached. It has to be taken in mind that for each applied shear the shearing time until equilibrium is reached is different. A higher shear rate would lead to more de-flocculation and thus to even lower viscosity at the equilibrium state. Still, the completely de-flocculated state which would be the ideal reference state for rheological measurements is not possible.

5 Conclusion

The experimental results show a good compliance between the performed investigations on cement pastes and the already existing theory of suspension agglomeration and colloidal particle interactions. For the rheological characterization of cementitious suspensions, it could be clarified that the preshear time is a determining factor on the experimental results. Theoretically, the ideal reference state for rheological measurements would be a completely de-flocculated system. This reference state however is not possible due to the instantaneous formation of particle networks after mixing. The most appropriate reference state for each introduced preshear energy should thus be figured out before rheological measurements. This is the time where an equilibrium of viscosity and yield stress are reached since this represents the most possible dispersed state of the suspension for that given shear load. It has to be taken in mind that with increasing preshear time, applied energy to the system might cause a change in the chemical reaction behavior of the cement pastes. Due to that, the most appropriate compromise should be clarified.

Furthermore, the results show that absolute measurement values of rheological parameters of cement pastes should not be compared if they are taken with noteworthy different experimental procedures, even if the same raw materials and mixture composition are used.

It can be pointed out that the state of de-flocculation of the system moreover has a decisive effect on the following structural build-up of cement pastes. This has to be taken into account for investigation on the thixotropy of cementitious suspensions. Experimental procedures regarding the effect of the preshearing on the thixotropic structural build-up of cementitious suspensions are currently being investigated.

Bibliography

[1] Larrard, F. de; Ferraris, C. F.; Sedran, T.: Fresh concrete: A Herschel-Bulkley material. In: Materials and Structures 31 (1998), S. 494–498

[2] Ferraris, Chiara F.: Measurement of the Rheological Properties of High Performance Concrete: State of the Art Report. In: Journal of Research of the National Institute of Standards and Technology 104 (1999), Nr. 5, S. 461–478

[3] Bentz, Dale P.; Ferraris, Chiara F.; Galler, Michael A.; Hansen, Andrew S.; Guynn, John M.: Influence of particle size distributions on yield stress and viscosity of cement–fly ash pastes. In: Cement and Concrete Research 42 (2012), Nr. 2, S. 404–409

[4] Snabre, P. ; Mills, P.: II. Rheology of Weakly Flocculated Suspensions of Viscoelastic Particles. In: Journal de Physique III 6 (1996), Nr. 12, S. 1835–1855

[5] Snabre, P.; Mills, P.: I. Rheology of Weakly Flocculated Suspensions of Rigid Particles. In: Journal de Physique III 6 (1996), Nr. 12, S. 1811–1834

[6] Bingham, Eugene C.: An Investigation of the Law of Plastic Flow. In: Bulletin of the Bureau of Standards 13 (1916), Nr. 266, S. 309–353

[7] Genovese, Diego B.: Shear rheology of hard-sphere, dispersed, and aggregated suspensions, and filler-matrix composites. In: Advances in colloid and interface science 171-172 (2012), S. 1–16

[8] Roussel, Nicolas: Rheology of fresh concrete : From measurements to predictions of casting processes. In: Materials and Structures 40 (2007), Nr. 10, S. 1001–1012

[9] Roussel, Nicolas; Lemaître, Anael; Flatt, Robert J.; Coussot, Philippe: Steady state flow of cement suspensions : A micromechanical state of the art. In: Cement and Concrete Research 40 (2010), Nr. 1, S. 77–84

[10] Plank, Johann; Hirsch, Christian: Impact of zeta potential of early cement hydration phases on superplasticizer adsorption. In: Cement and Concrete Research 37 (2007), Nr. 4, S. 537–542

[11] Flatt, Robert J.; Bowen, Paul: Electrostatic repulsion between particles in cement suspensions : Domain of validity of linearized Poisson–Boltzmann equation for nonideal electrolytes. In: Cement and Concrete Research 33 (2003), Nr. 6, S. 781–791

[12] Flatt, Robert J.; Bowen, Paul; Houst, Y. F. ; Hofmann, H.: Modelling interparticle forces and yield stress of cement suspensions. In: Proceedings of the 11th International Congress on the Chemistry of Cement (ICCC), S. 618–627

[13] Malkin, A.Ya.: Non-Newtonian viscosity in steady-state shear flows. In: Journal of Non-Newtonian Fluid Mechanics 192 (2013), S. 48–65

[14] Barnes, Howard A.; Hutton, John F.; Walters, Kenneth: An introduction to rheology. 3. impression. Amsterdam: Elsevier, 1993 (Rheology series 3)

[15] Flatt, Robert J. ; Bowen, Paul: Yodel : A Yield Stress Model for Suspensions. In: Journal of the American Ceramic Society 89 (2006), Nr. 4, S. 1244–1256

[16] Struble, Leslie; Sun, Guo-Kuang: Viscosity of Portland cement paste as a function of concentration. In: Advanced Cement Based Materials 2 (1995), Nr. 2, S. 62–69

[17] Roussel, N.; Ovarlez, G.; Garrault, S.; Brumaud, C.: The origins of thixotropy of fresh cement pastes. In: Cement and Concrete Research 42 (2012), Nr. 1, S. 148–157

[18] D. C-H. Cheng: Yield stress: A time-dependent property and how to measure it. In: Rheologica Acta (1986), Nr. 25, S. 542–554

[19] Mehdipour, Iman ; Khayat, Kamal H.: Understanding the role of particle packing characteristics in rheo-physical properties of cementitious suspensions : A literature review. In: Construction and Building Materials 161 (2018), S. 340–353

[20] Mujumdar, Ashutosh; Beris, Antony N.; Metzner, Arthur B.: Transient phenomena in thixotropic systems. In: Journal of Non-Newtonian Fluid Mechanics 102 (2002), Nr. 2, S. 157–178

[21] Sonebi, M.; Lachemi, M.; Hossain, K.M.A.: Optimisation of rheological parameters and mechanical properties of superplasticised cement grouts containing metakaolin and viscosity modifying admixture. In: Construction and Building Materials 38 (2013), S. 126–138

[22] Ferrari, Lucia; Kaufmann, Josef; Winnefeld, Frank; Plank, Johann: Interaction of cement model systems with superplasticizers investigated by atomic

force microscopy, zeta potential, and adsorption measurements. In: Journal of Colloid and Interface Science 347 (2010), Nr. 1, S. 15–24

[23] Zhang, Yanrong: Study on Microstructure and Rheological Properties of Cement-Chemical Admixtures-Water Dispersion System at Early Stage. Peking, Tsinghua University, Department of Civil Engineering. Dissertation. 2017

[24] Dougherty, Thomas J.; Krieger, Irvin M.: Potential Around a Charged Colloidal Sphere 63 (1959), S. 1869–1872

[25] Barnes, Howard A.: The yield stress - a review or panta rei - everything flows? . In: Journal of Non-Newtonian Fluid Mechanics 81 (1999), S. 133–178

[26] Barnes, Howard A.: Thixotropy—a review. In: Journal of Non-Newtonian Fluid Mechanics 70 (1997), 1-2, S. 1–33

[27] Roussel, Nicolas: A thixotropy model for fresh fluid concretes : Theory, validation and applications. In: Cement and Concrete Research 36 (2006), Nr. 10, S. 1797–1806

[28] Goodeve, C. F.: General theory of thixotropy and viscosity (1938)

[29] Roussel, N.; Stefani, C.; Leroy, R.: From mini-cone test to Abrams cone test : Measurement of cement-based materials yield stress using slump tests. In: Cement and Concrete Research 35 (2005), Nr. 5, S. 817–822

Advanced Admixtures to control Rheology of Tremie Concrete for Deep Foundations

O. Mazanec[1)], S. Dittmar[2)], P. Gaeberlein[3)]
[1)] BASF Construction Solution GmbH, Trostberg, Germany
[2)] BASF Construction Solution GmbH, Mannheim, Germany
[3)] BASF Schweiz AG, Holderbank, Switzerland

Abstract

With increasing fluidity of concretes, the tendency of aggregates to sediment and separate increases. The resulting failure modes can range from inferior surface appearance to reduced structural strength. In modern concrete, the sedimentation and water loss can be prevented by the addition of viscosity modifying admixtures (VMA). In most cases VMAs are based on modified polysaccharides or synthetic copolymers.

First the viscosity enhancing effect of the different VMAs in aqueous solutions and its evolution around the so-called critical aggregation concentration has been studied. To determine the rheological properties, rheometer measurements of the aqueous solutions modified with the different polymers were performed at several polymer dosages. The rheological behaviour was measured using an Anton Paar Rheometer MCR 302 equipped with coaxial cylinder sensor system. It was found that some aqueous VMA solution exhibits at low dosages a Newtonian region and a shear thinning region while others showed at low dosages only a Newtonian region. We moreover measured the existence of a critical aggregation concentration at which the macromolecule chains of the polymers interact which each other. The critical aggregation concentration of VMAs with associative side chains is much lower than that of VMAs without associative side chains. At dosages above the critical aggregate concentration polymer interaction take place which lead to the shear thinning behaviour.

Finally, we could transfer the results from aqueous solution measurements to the macroscopic properties of highly flowable tremie concrete for deep foundations, where long workability and high water retention properties are required. The different rheological profiles of the VMAs effect the pumpability, sedimentation and water retention properties of the concrete. It was found that VMAs with constant properties at different shear rates (Newtonian region) are beneficial to stabilize against water loss under high pressure, whereas VMAs with shear thinning behaviour are highly efficient to stabilize highly flowable concrete against sedimentation and bleeding.

Keywords: viscosity modifying admixtures, aggregate segregation, water retention, bleeding, stabilizing polymer

Analysis of the effect of various limestones in the cement composition on rheological properties of mortar

Jacek Gołaszewski, Małgorzata Gołaszewska
Silesian University of Technology, Gliwice, Poland
e-mail: malgorzata.golaszewska@polsl.pl

1 Introduction

Addition of limestone as a constituent of cement can be beneficial for properties of mortars, as it can lead to a lower water demand, better workability, and an increase in the durability of concrete, when in combination with other mineral additives [1,2]. However, the influence of limestone as a constituent of cement on the rheological properties of mortars, can significantly vary, depending on the limestone used [3,4]. Presented research was conducted in order to test the rheological properties of mortars with cement with six limestones of different origins. Tests included measurement of yield stress and plastic viscosity, and mortar consistency.

2 Materials and measurement methods

Limestone cements were obtained by mixing Portland cement CEM I 42,5R with six different limestones in amounts of 15, 30 and 40%.

All measurements were conducted on mortars which composition was based on standard mortar according to EN-197 – 1 [5], however due to technical aspects, w/c ratio was raised to 0.55. Mortars have been subjected to rheological tests in a rheometer Viskomat NT and consistency test according to standard PN-B-04500 [6] after 5 and 60 minutes from mixing. To calculate the yield stress and plastic viscosity of the mortar, simplified Bingham model was adopted.

3 Results

Obtained results have shown different effect of addition of limestone to cement in amount of 15-40% on yield stress of mortars. In case of one limestone, yield stress increased with the increase of limestone content, however generally, limestone addition reduced the yield stress of mortar. It should be noted, that addition of 15% of limestone did not significantly influence the yield stress after both 5 and 60 min.

Plastic viscosity of cement with limestone was similar or higher than in the case of Portland cement and generally increased with the increase in limestone content with cement.

Consistency of limestone cements corresponds to yield stress – the mortar with limestone cement that was characterized by increased yield stress had visibly lower flow diameter than reference cement, while in case of other limestones, the flow diameter is similar or larger.

The difference in behavior of limestones could be related to their particle size distribution, as the limestone which addition increased yield stress was characterized by bimodal, and thus discontinuous, particle size distribution.

4 Conclusions

Conducted tests indicate, that the type of limestone can significantly influence the rheological properties of mortars. The difference in effect of limestone addition was linked to discontinuous particle size distribution.

Bibliography

[1] Boos, P., Haerdtl, R.; Experience report Portland limestone cement, Heidelberg Cement Group, 2004.

[2] Schmidt M.: Cement with interground additives, Zement-Kalk-Gips, 87–92 1999.

[3] G. Bolte, Zemente mit hohem Kalksteingehalt, 19. Int. Baustofftagung, 405–415, Fischer H.-B., Weimar: Bauhaus-Universität Weimar, 2018.

[4] Courard L., Herfort D., Villagran Y.: Performances of limestone modified portland cement and concrete, Technical Report, RILEM, 2016.

[5] EN 196-1:2016: Methods of testing cement. Determination of strength

[6] PN-85-B-04500 Zaprawy budowlane-Badanie cech fizycznych

Investigations on the influence of paste composition on the rheological properties of flowable mortars and concretes

Wibke Hermerschmidt, Jaqueline Honecker, Christoph Müller
VDZ gGmbH, Düsseldorf, Germany
Phone: +49 211-4578-1, e-mail: wibke.hermerschmidt@vdz-online.de

Abstract

The presented study investigates effects of the paste composition on the rheological properties of mortars and on the consistency of concrete. The parameters being varied are the equivalent water/cement ratio, the cement type and the dosage of superplasticizer. The yield stress and the plastic viscosity of mortars are derived from rheological measurements and analyzed regarding influences of the above-mentioned parameters as well as correlations to the slump flow. Furthermore, it is checked if there are qualitative correlations between the fresh mortar and concrete properties in case of identical paste compositions.

1 Introduction

The workability of flowable mortars and concretes is mainly determined by their rheological properties, i.e. yield stress and plastic viscosity. Therefore, a rheological characterization of mortars and concretes can be useful on the one hand to optimize the workability and on the other hand to choose the right way of placing concrete on site with regard to sufficient compaction, minimized segregation and bleeding etc.

The objective of this contribution is to study effects of the paste composition on the rheological properties of mortars and on the consistency of concrete. Rheological measurements and slump flow tests were carried out to characterize the workability of mortars with different paste compositions. The compositions were systematically varied regarding the equivalent water/cement ratio, the cement type and the dosage of superplasticizer. As a second step, the compositions of the mortars were scaled up to concrete compositions with a max. aggregate size of 16 mm. The obtained results regarding the concrete consistency were used to identify correlations between the concrete and mortar properties.

2 Experimental

2.1 Materials

Four different cements – one Portland cement, one Portland-slag cement and two blast furnace cements with different contents of granulated blast furnace slag (GGBFS) – were used for the concrete and mortar tests. The relevant properties of the cements are listed in Table 1. Silicious fly ash from a German power plant was used as type II addition and considered in the mix design with a k-value of 0.4. The sand and gravel used in the mortars and concretes are standard aggregates from the Rhine river. A polycarboxylate ether (PCE) superplasticizer was used to adjust the flowability of mortars and concretes.

Table 1: Cement types and properties

Cement	Cement type	GGBFS [1)] [wt%]	Fineness [2)] [cm²/g]	Water demand [3)] [wt%]
ZI	CEM I 42,5 N	-	3050	25.5
ZII	CEM II/A-S 42,5 R	16	3300	27.5
ZIII	CEM III/A 42,5 N	45	3690	31.0
ZIV	CEM III/B 42,5 L	69	4270	32.5

1) Content of blast furnace slag determined acc. to CEN/TR 194-4
2) Fineness of cement determined acc. to EN 196-6, Blaine method
3) Water content for standard consistency determined acc. to EN 196-3

2.2 Mixture proportions

Table 2 gives an overview on the compositions of the investigated concretes. The three concretes differ in equivalent water/cement ratio $(w/c)_{eq}$ while the paste volume (i.e. the sum of volumes of cement, fly ash and water), the ratio between cement content and fly ash content and the grading curve of the aggregates were kept constant for all concretes. The four mortar compositions given in Table 3 follow the same principle. The volume ratios of paste to sand and cement to fly ash are similar within the mortars and the concretes. This means that for a defined $(w/c)_{eq}$-value the mortar composition is equivalent to the corresponding concrete composition without coarse aggregates.

The three concrete mixes were prepared using the four cements listed in Table 1, which leads to a total number of 12 test series. The same procedure was applied for the mortar tests, which leads to a total number of 16 test series. Within one test series, the dosage of superplasticizer was varied to adjust the flowability of mortars and concretes.

The mortars were mixed in a standard Hobart mixer acc. to the procedure given in EN 196-1. The concretes were mixed in laboratory mixers with a nominal volume of 30 l and 75 l respectively. Cement, fly ash and aggregates were mixed for 30 s. Water and superplasticizer were added steadily within a period of approximately 30 s while the mixer was running. Mixing was continued for 120 s.

Table 2: Mixture proportions of concretes

Concrete mix	Cement [kg/m³]	Fly ash [kg/m³]	$(w/c)_{eq}$ [-]	Superplast. [wt% of cem.]	Aggregates	Paste volume [l/m³]
C-0.30	395	110	0.30	1.2 … 1.5	river sand and gravel, A/B16	≈ 320 l/m³
C-0.46	320	90	0.46	0.4 … 0.8		
C-0.75	240	68	0.75	0.0 … 0.3		

Table 3: Mixture proportions of mortars

Mortar mix	Cement [kg/m³]	Fly ash [kg/m³]	$(w/c)_{eq}$ [-]	Superplast. [wt% of cem.]	Aggregates	Paste volume [l/m³]
M-0.30	755	210	0.30	0.6 … 1.5	river sand 0/2 mm	≈ 615 l/m³
M-0.46	620	175	0.46	0.0 … 0.4		
M-0.60	530	150	0.60	0.0 … 0.3		
M-0.75	460	130	0.75	0.0 … 0.2		

2.3 Test methods

The consistency of mortars was tested with a flow table test referring to EN 1015-3. In contrast to the procedure described in EN 1015-3, the flow table was not lifted and dropped, meaning the spread was caused solely by the flowability of the mortar under self-weight without external impacts. The test was started approximately 5 min after the mixing procedure was finished.

The consistency of concretes was tested with the flow table test acc. to EN 12350-5. In addition, the flow time after raising the mould until the concrete stops to flow by itself was measured together with the flow table spread before lifting and dropping of the flow table. The test was started approximately 15 min after the mixing procedure was finished. Directly before, the concrete was remixed for 30 s.

The rheological properties of mortars were investigated using a rotational rheometer (type "Viskomat NT" of Schleibinger) equipped with a vane-probe. The test vessel with an inner diameter of 100 mm and inner height of 110 mm rotated with a defined speed around a steel vane with 6 uniformly placed blades, a diameter of 40 mm and a height of 60 mm. The rotational speed was defined according to the speed-time profile shown in Fig. 1. The torque value measured at the end of each speed plateau was used to calculate the yield stress and the plastic viscosity of the mortars. For the calculation a Bingham constitutive model was assumed. The geometry of the vessel with the vane-probe was approximated by two concentric cylinders. The yield stress and the plastic viscosity were calculated based on the "effective annulus method" described in [1].

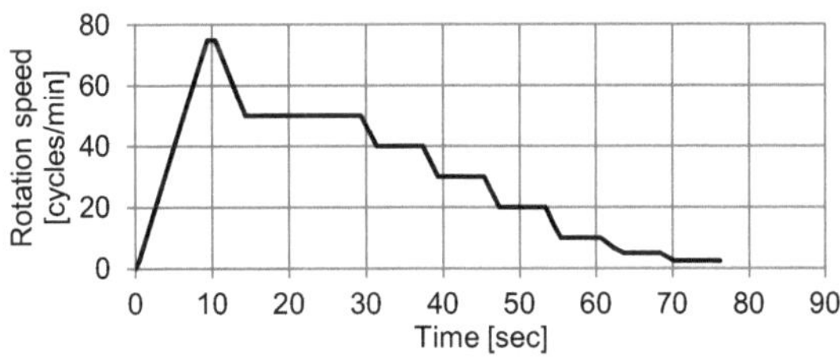

Figure 1: Speed-time profile for rheological measurements

3 Results and discussion

3.1 Mortar properties

The slump flow values of the mortars are shown in Fig. 1 as a function of the dosage of superplasticizer (SP-dosage). As expected, the lower the equivalent water/cement ratio of the mortar, the higher the necessary SP-dosage for a defined slump flow. The influence of the cement type on the necessary SP-dosage for a defined slump flow is more pronounced, the lower the equivalent water/cement ratio. The slump flow of the mortars M-0.60 and M-0.75 respectively shows a relatively small disparity when the SP-dosage is kept constant and the cement is varied. In contrast the slump flows of the mortars M-0.30 and M-0.46 show a clear dependency between the cement type and the demand of SP for a defined flow. The comparison between the mortars M-0.30-ZII, M-0.30-ZIII, M-0.30-ZIV and M-0.46-ZII, M-0.46-ZIII, M-0.46-ZIV reveals that the SP-dosage necessary for a defined slump flow at constant $(w/c)_{eq}$ decreases with increasing GGBFS content of the cement. Herrmann & Rickert [2] and Palacios et al. [3], among others, reported comparable effects for cement pastes and mortars respectively. Acc. to [2] and [3], the effects are linked to a different PCE adsorption behavior of clinker and GGBFS and to the composition of the pore solution that varies in dependence to the content of GGBFS. According to these explanations the Portland cement ZI would need the highest SP-dosage to reach a given flow value, which is not the case. This is probably due to the fact that the Portland cement ZI comes from another cement plant than the (Portland-)slag cements ZII, ZIII and ZIV. Therefore, it has a different clinker composition that probably leads to a different interaction with the superplasticizer compared to the clinker component of the (Portland-)slag cements.

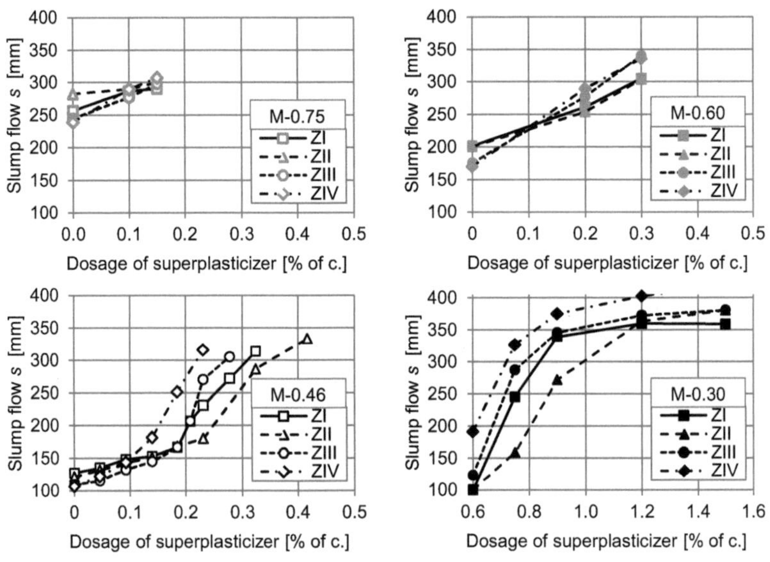

Figure 2: Slump flow of mortars as a function of the dosage of superplasticizer

Fig. 2 shows the yield stress and the plastic viscosity of the mortars derived from the rheological tests. With exception of the mortars M-0.30 there is a good correlation between the yield stress and the slump flow that seems to have no significant dependence on the equivalent water/cement ratio and the cement type. The yield stress decreases approximately exponentially with increasing slump flow. A possible explanation for the fact that the mortars M-0.30 do not match this correlation is the thixotropic behavior of the mortars that is more pronounced for low equivalent water/cement ratios and implies dependence between the yield stress and the shear history. This leads to the fact that the yield stress derived from the rheological measurement is only an apparent value because the Bingham model does not take into account any effects of thixotropy.

The plastic viscosity shows a decreasing trend with increasing slump flow. If mortars with a similar slump flow are compared, the equivalent water/cement ratio is the most decisive parameter influencing the plastic viscosity. A low $(w/c)_{eq}$ is connected with a high viscosity and vice versa. For the mortars M-0.30 there is no clear influence of the cement type on the viscosity. The same applies to the mortars M-0.46 and M-0.60 with slump flow values $s < 200$ mm. For the remaining mortars the Portland cement ZI leads to a lower plastic viscosity compared to the other cements while the differences between the mortars prepared with the slag cements ZII, ZIII and ZIV are relatively small.

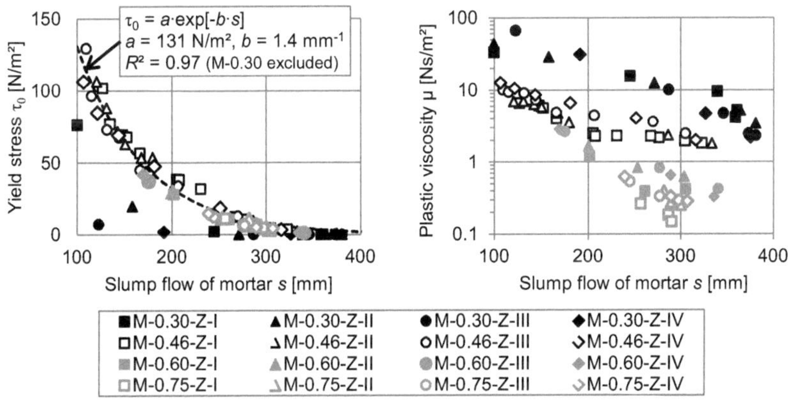

Figure 3: Yield stress (left) and plastic viscosity (right) of mortars

3.2 Concrete properties

The flow values of the concretes with different dosages of superplasticizer are shown in Fig. 3 (left). The flow values of the concretes prepared with the (Portland-)slag cements (ZII, ZIII and ZIV) were qualitatively comparable to the slump flow values of the mortars, i.e. the flow values for a defined SP-dosage increased with increasing GGBFS content of the cement. In contrast to the mortars, the concretes prepared with the Portland cement ZI showed the highest SP-demand. This might be due to the fact that the flow table tests for the concretes were carried out 15 minutes after mixing while the slump flow tests for the mortars were carried out approximately 5 minutes after mixing. Thus, the comparison is impacted by the setting behavior of the concretes which in turn depends on the characteristics of the cements.

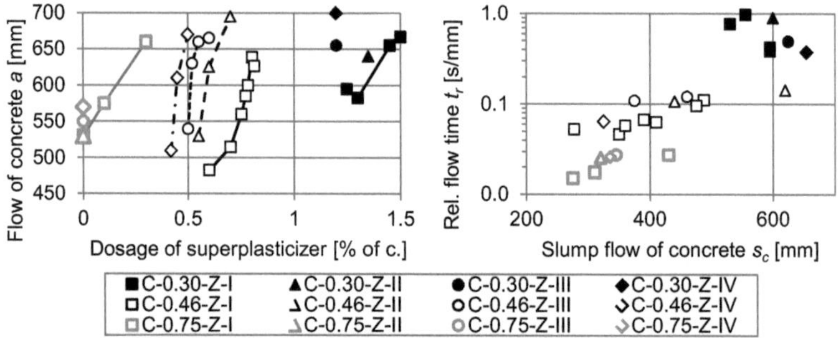

Figure 4: Flow of concretes as a function of the dosage of superplasticizer (left); Relative flow times of concretes (right)

The flow time t_{sc} in the flow table test, meaning the time from raising the mould until the concrete stops to flow by itself, was analyzed to get a qualitative assessment of the viscosity of the concretes. Because the flow time depends on the slump flow s_c (spread diameter before lifting the table), a relative flow time t_r was calculated acc. to equation 1:

$$t_r = \frac{2\,t_{sc}}{s_c - 200\,\mathrm{mm}} \tag{1}$$

The relative flow time t_r can be interpreted as mean time necessary to increase the radius of the slump flow by 1 mm. The values of t_r are shown in Fig. 3 (right). In comparison to the plastic viscosity of the mortars (Fig. 2) a similar categorization by equivalent water/cement ratio was found for the relative flow times of the concretes. This indicates that the determination of the relative flow time in the flow table test can be a useful tool to categorize the viscosity of concretes. Thus, it might be an effective way to obtain additional information on the workability of concrete without much extra effort.

4 Conclusions

The effects of the paste composition on the rheological properties of mortars and on the consistency of concretes were investigated in the presented study. The equivalent water/cement ratio $(w/c)_{eq}$, the cement type and the dosage of superplasticizer were the governing parameters that were varied within the study. The results showed that the demand of superplasticizer (PCE) for a given flow of mortar and concrete respectively depends both on the equivalent water/cement ratio and the cement type. The slump flow of the mortars with $(w/c)_{eq} \geq 0.46$ correlated with the yield stress derived from the rheological measurements. This correlation was found to have no significant dependence on the paste composition. With regard to the viscosity, the equivalent water/cement ratio was found to be the most relevant parameter while the impact of the cement type was comparatively low. Most of the findings from the mortar tests correlated qualitatively with the results of the concrete tests if mortars and concretes with identical paste compositions were compared. Furthermore, it has been shown that measuring the flow time of concrete as an additional value in the flow table test acc. to EN 12350-5 might be a useful tool to categorize the viscosity of very soft or flowable concretes.

7 Acknowledgements

This research is part of the IGF project 19276 N, which is funded by the Federal Ministry for Economic Affairs and Energy via the AiF as part of the IGF-program on the basis of a resolution of the German Bundestag.

Bibliography

[1] Koehler, E. P.; Fowler, D. W.: Development of a portable rheometer for fresh portland cement concrete, Research Report, Austin, 2004.

[2] Rickert, J.; Herrmann, J.: Interactions between granulated blastfurnace slag or limestone as cement main constituent and super-pasticisers based on polycarboxylate ether. In: 13th International Congress on the Chemistry of Cement, ICCC (Madrid 03.–08.07.2011), 2011

[3] Palacios, M., et al.: Effect of PCs superplasticizers on the rheological properties and hydration process of slag-blended cement pastes. Journal of Materials Science 44(10), pp. 2714-2723, 2009

Rotatorische und oszillatorische Scherversuche zur Ermittlung steifigkeitsrelevanter Kenngrößen von Offshore-Vergussmörteln unter dem Einfluss des Early-Age Movement

Dario Cotardo, Prof. Michael Haist, Prof. Ludger Lohaus, Christoph Begemann
Institut für Baustoffe der Leibniz Universität Hannover

Kurzfassung

Relative movements, prior to and during the initial setting of the grout are to be regarded as particularly critical for the structural soundness of grouted connections. Currently, the expected relative movement during grouting shall be limited to a maximum of 1 mm, relying on measurements on ungrouted piles. However, it can be assumed that the grout contributes to the stiffness of the structure in a significant manner. In order to consider the grout-stiffness during the design process, methods for determining the development of grout-stiffness under the influence of early-age movement are required. The goal of the presented research was to develop such a method, which is based on rotational and oscillatory rheology and will be presented in the following. [5]

1 Einleitung

Tragstrukturen von Offshore-Anlagen (Windenergieanlagen und Umspannplattformen) sind beginnend mit der Installationsphase Wind- und Wellenbelastungen ausgesetzt. Besonderes Augenmerk muss auf den Übergang zwischen den Pfählen und der Tragstruktur gerichtet werden, der üblicherweise als Rohr-in-Rohr-Steckverbindung, als so genannte Grout-Verbindung realisiert wird. Der Ringspalt zwischen den beiden Stahlrohrelementen wird mit einem zementgebundenen Baustoff, dem Grout-Material, gefüllt. Das Grout-Material führt nach seiner Erhärtung zu einer kraft- und formschlüssigen Verbindung der beiden Stahlrohrelemente. Während des Verfüll- und Aushärteprozesses des Grout-Materials werden Grout-Verbindungen wellenbedingten Relativbewegungen ausgesetzt (vgl. Bild 1). Dies geschieht bereits bei geringen Wellenhöhen und dementsprechend bei sehr häufig vorkommenden Wetterbedingungen.

Relativbewegungen während der Erhärtung des Grout-Materials werden als besonders kritisch bewertet. In dieser Phase können die sich ausbildenden Steifigkeits- und Festigkeitseigenschaften des Materials durch die Bewegungen gestört und dauerhaft beeinträchtigt werden. Relativbewegungen zwischen den Stahlrohrelementen während der Erstarrung und Erhärtung des Grout-Materials werden als Early-Age Movement (EAM) bezeichnet. Auch wenn das EAM in aktuellen Normen und Richtlinien [8][9] explizit angesprochen wird, ist nahezu unklar, wie diesem Phänomen begegnet werden soll.

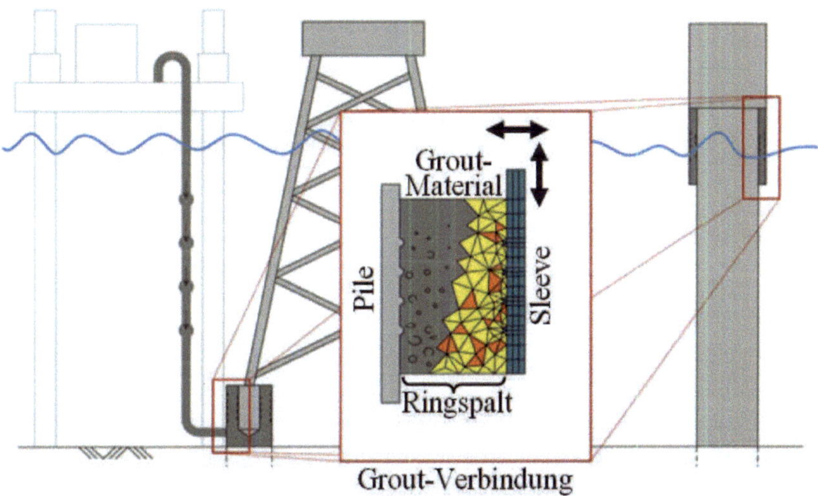

Bild 1: Schematische Darstellung des Early-age Movement während der Installationsphase einer Jacket-Tragstruktur (links) und eines Monopiles (rechts). [5]

2 Stand der Forschung

Zum Einfluss des EAM auf die Tragfähigkeit von Grout-Verbindungen, gibt es nur wenige Erkenntnisse, die auf die 1980er und 1990er Jahre zurückgehen [3][13][15]. Die Ergebnisse sind jedoch nur teilweise auf heutige Offshore-Anlagen übertragbar. Aktuelle Untersuchungen zeigen, dass welleninduzierte Relativverschiebungen im frühen Alter die Materialeigenschaften beeinflussen können [17][18]. Es wurde beobachtet, dass solche Relativverschiebungen zu Entmischungsprozessen im Material führen können.

Darauf aufbauende Untersuchungen haben gezeigt, dass das EAM auch zu einer verstärkten Entlüftung des Grout-Materials führen kann, was die Druckfestigkeit erhöht (vgl. Bild 2). In Untersuchungen von [19] wurde der Einfluss von Relativverschiebungen betrachtet, die auf Grundlage einer numerischen Simulation ermittelt wurden. Dabei wurde bislang stets eine nicht vergossene Verbindung berücksichtigt, so wie es in einschlägigen Normen (vgl. z.B. [8]) vorausgesetzt wird.

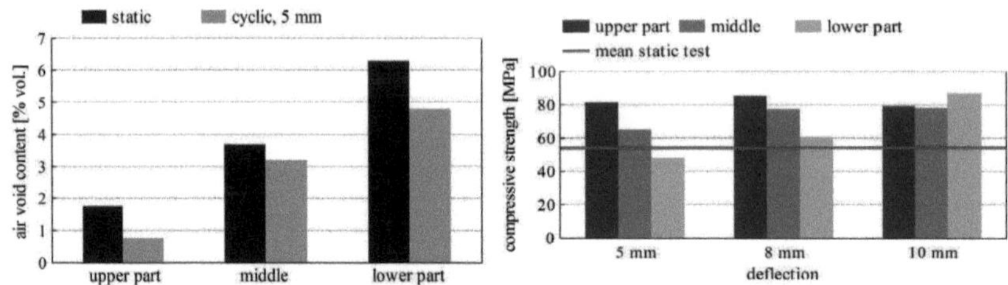

Bild 2: Luftgehalt (links) und Druckfestigkeit (rechts) von erhärteten Proben in Abhängigkeit ihrer Lage in einer nachgestellten Grout-Verbindung mit und ohne den Einfluss des EAM. [19][20]

Derzeit fehlen entsprechende Ansätze, die eine vergossene Grout-Verbindung berücksichtigen. Unter der Annahme, dass bereits das frische Grout-Material zu einer Erhöhung der Steifigkeit der Tragstruktur beiträgt und sich dadurch der Widerstand gegenüber Wind- und Wellenbelastungen erhöht, werden sich geringere Relativverschiebungen einstellen. Dazu ist es erforderlich, steifigkeitsrelevante Materialkennwerte während der Frischmörtelphase bis zur Erstarrung des Grout-Materials zu bestimmen. Diese können dann wiederum als Eingangsgrößen in numerische Simulationen einfließen.

Zeitabhängige Materialeigenschaften können mit oszillatorischen und rotatorischen rheologischen Messungen [24][28] und zusätzlich mit Ultraschallverfahren [16][26][27] bestimmt werden. Allerdings eignen sich Ultraschallmessungen vorwiegend für eine spätere Phase der Erstarrung. Für die Bestimmung der Materialeigenschaften in der frühen Phase bis zur Erstarrung sind rheologische Messungen besser geeignet. Grundlagen für rheologische Messungen finden sich beispielsweise in [23]. Wesentliche rheologische Materialkennwerte für Grout-Materialien sind die Fließgrenze τ_0 und die Viskosität μ, die vom BINGHAM-Modell abgeleitet werden können (vgl. Gl. 1 und Bild 3, links).

In der Regel werden diese Kenngrößen durch rotatorische rheologische Messungen bestimmt. Dabei wird die resultierende Scherspannung τ durch Steuerung der Scherrate $\dot{\gamma}$ gemessen und anschließend als Funktion von der Scher-rate $\dot{\gamma}$ aufgetragen. Anhand der auf diese Weise resultierenden Fließkurven (vgl. Bild 3) können die Fließgrenze τ_0 und die plastische Viskosität μ bestimmt werden. Das BINGHAM-Modell kann weder scherverdickendes noch scherverdünnendes Verhalten beschreiben, weshalb für zementgebundene Baustoffe, häufig das HERSCHEL-BULKLEY-Modells angewendet wird ([6]; vgl. Gl.2).

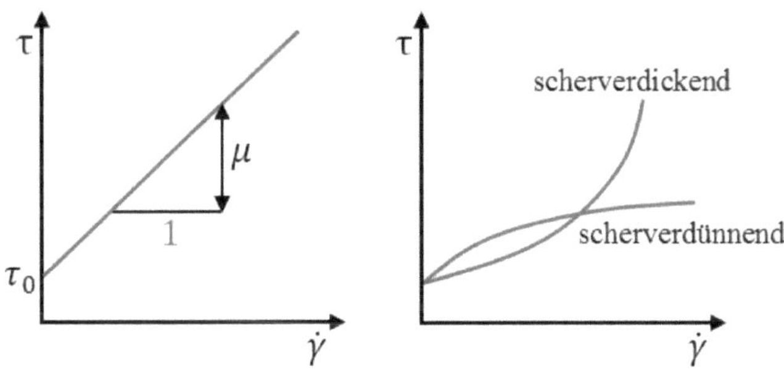

Bild 3: Fließkurve eines Bingham-Fluids (links) und scherverdickendes und scherverdünnendes Fließverhalten mit Fließgrenze (rechts). [5]

$$\tau(\dot{\gamma},t) = \tau_0(t) + \mu(t) \cdot \dot{\gamma} \quad (1)$$

$$\tau(\dot{\gamma},t) = \tau_0(t) + \mu(t) \cdot \dot{\gamma}^{p(t)} \quad (2)$$

In Gl. 2 bezeichnet p den dimensionslosen HERSCHEL-BULKLEY-Index. Um die Entwicklung der Materialeigenschaften zu untersuchen, müssen drei Parameter (τ_0, μ und p) bestimmt werden. Da zementgebundene Materialien eine ausgeprägte chemische Reaktivität besitzen, hängen alle drei Parameter vom Alter der Probe und damit von der Zeit t ab. Darüber hinaus beeinflussen physikalische Wechselwirkungen zwischen den Partikeln zusätzlich die Eigenschaften des Grout-Materials. Im Folgenden wird daher zwischen chemisch induzierten Strukturänderungen (d.h. eine Zunahme der Struktur und damit Steifigkeit) und Agglomerations- und Dispersions-induzierten Strukturänderungen unterschieden.

Beide Effekte beeinflussen die rheologischen Eigenschaften des Grout-Materials. Um sie zu quantifizieren, ist es zwingend erforderlich, die Proben über ein breites Spektrum an Scherraten $\dot{\gamma}$ zu scheren. Dabei wird jedoch die Struktur der Substanz gestört, d.h. ein Strukturbruch tritt ein. Ein solcher Strukturbruch führt zum Aufbrechen von Agglomeratstrukturen (Dispergierung), die durch Hydratation oder kolloidale Wechselwirkungskräfte entstanden sind [7][29][30][31]. Dabei kann ein Rückgang in der messbaren Fließgrenze τ_0 und plastischen Viskosität μ wahrgenommen werden, was Entmischungsphänomene, z.B. Sedimentationserscheinungen [4][21], begünstigen kann. Neben dem durch Scherung initiierten Strukturbruch kommt es darüber hinaus zur Orientierung ungleichförmiger Partikel in Fließrichtung [2][25], was die Fließgrenze τ_0 und die Viskosität μ maßgeblich verringert, wie in Bild 4 schematisch dargestellt ist.

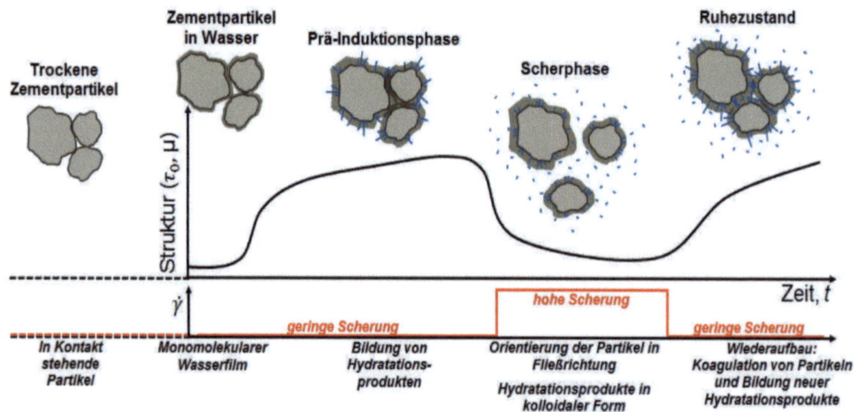

Bild 4: Schematische Darstellung eines Strukturaufbaus, Strukturbruchs und Wiederaufbaus einer zementgebundenen Suspension in Abhängigkeit der Schereinwirkung. [2][5]

Neben einem scherinduzierten Strukturbruch können geringe Scherspannungen auch zu einem Strukturaufbau führen, da durch den Energieeintrag repulsive Kräfte, die zwischen zwei interagierenden Partikeln herrschen, überwunden werden. Dies führt zur Agglomeration der Partikel. Grundsätzlich kann die Struktur frischer Zementsuspensionen durch Koagulationszustände der Partikel als eine Funktion von der Scherbeanspruchung beschrieben werden.

Agglomeratstrukturen können durch einen scherbedingten Impulsaustausch (extrinsische Agglomeration) gebildet werden. Dabei stellt sich ein belastungsspezifischer Gleichgewichtszustand zwischen strukturaufbauenden und strukturabbauenden Prozessen ein [12][14]. Die mechanischen Eigenschaften dieser Agglomeratstrukturen können mit oszillatorischen Scherversuchen (vgl. Bild 5) erfasst werden, bei denen die Probe nur geringen Belastungen für sehr kurze Zeitspannen ausgesetzt wird. Ferner erlaubt dieses Verfahren die Bestimmung zeitvarianter rheologischer Materialeigenschaften, z.B. zur Bestimmung der Thixotropie [22].

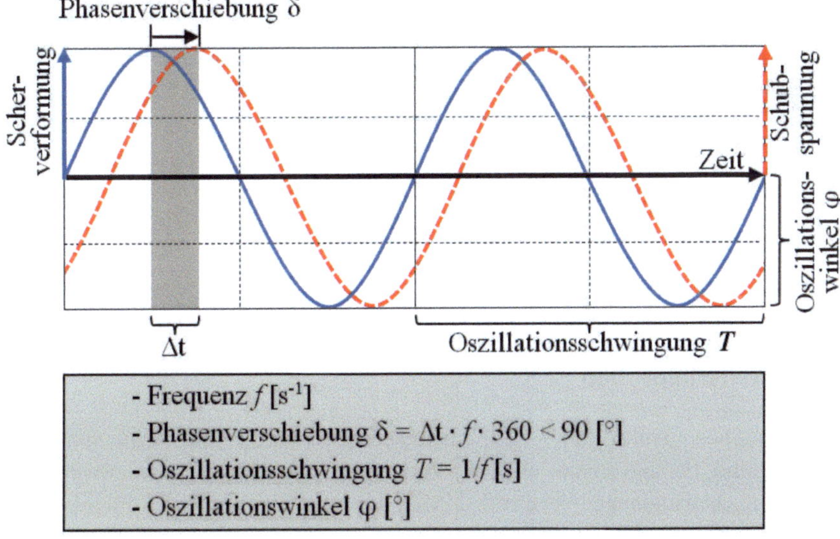

Bild 5: Exemplarische Darstellung eines oszillatorischen Scherversuchs einer zementgebundenen Suspension unter Vorgabe des Oszillationswinkels (Scherverformung). [5]

Das Modell ist jedoch nur anwendbar, wenn die Einwirkung die Fließgrenze τ_0 nicht überschreitet. Für die in diesem Artikel vorgestellten experimentellen Untersuchungen ist es jedoch beabsichtigt, das Material ins Fließen zu bringen ($\tau_0 \ll \tau$), um das EAM durch Oszillation zu simulieren. Aus diesem Grund werden oszillatorische Scherversuche nicht genutzt, um Kenngrößen zum elastischen Dehnungsverhalten des Grout-Materials zur Abbildung der Steifigkeitsentwicklung im Zeitverlauf zu bestimmt.

Die Oszillation wird in diesem Beitrag genutzt, um einerseits das EAM zu simulieren und andererseits, um über die Drehmomentamplitude T_A ausgewählter

Oszillationsschwingungen Auskunft über die Steifigkeitszunahme im Zeitverlauf zu erlangen.

3 Zielstellung

Ziel der Untersuchungen war es, eine Methode zur Bestimmung steifigkeitsrelevanter Materialkennwerte von Grout-Materialien unter dem Einfluss des EAM zu entwickeln. Zu diesem Zweck wurde das EAM mittels Oszillation simuliert. Darüber hinaus wurde die Steifigkeitsentwicklung des Grout-Materials durch Oszillations- und Rotationsrheologie abgebildet. Ziel war es, den Einfluss unterschiedlicher Relativverschiebungen auf die Steifigkeitsentwicklung eines Grout-Materials unter Berücksichtigung variierender Wasser/Feststoff-Verhältnisse zu quantifizieren. Die Steifigkeitsentwicklung wurde solange bestimmt, wie die Probe eine rheologische Messung zuließ (bis zu 430 min nach Wasserzugabe).

4 Materialien und Methoden

4.1 Materialproben

Ursprünglich sollte ein Hochleistungsvergussmörtel als Untersuchungsgegenstand herangezogen werden. Dieses in Deutschland üblicherweise für Offshore-Anwendungen verwendete Material enthält in der Regel granulare Zuschlagstoffe unterschiedlicher Durchmesser. Voruntersuchungen haben gezeigt, dass eine übermäßige Sedimentation während der rheologischen Messung insbesondere bei hohen Fließfähigkeiten auftreten kann, was die Ergebnisse verfälscht und auch zu einem vorzeitigen Abbruch der Messung führt. Dementsprechend ist ein Material grundsätzlich besser geeignet, das keine Gesteinskörnung enthält.

Aus diesem Grund wurde für die Untersuchungen ein Injektionsleim verwendet, der in der Zusammensetzung des Bindemittels dem ursprünglich angedachten Vergussmörtel sehr ähnlich ist und vom selben Hersteller bereitgestellt wurde.

Obwohl das Gesteinskorn zur Steifigkeit des Materials beiträgt, trägt es nicht zur zeitabhängigen Entwicklung der Steifigkeit bei, sodass das Material zur Entwicklung der in diesem Artikel beschriebenen Prüfmethode geeignet ist. Zur Herstellung der Materialproben wurde das Material nach Wasserzugabe für eine

Zeitdauer von 8 min gemischt. Die Probenpräparation erfolgte direkt nach dem Mischprozess. Variierende Fließfähigkeiten wurden durch Anpassung des Wasser/Feststoff-Verhältnisses (w/f-Wert) eingestellt. Die Fließfähigkeit wurde mit dem Haegermann-Konus (Ausbreitfließversuch) nach DIN EN 1015-3 bestimmt. Das Ausbreitfließmaß wurde auf 150 mm (Materialprobe A), 300 mm (Materialprobe B) und 450 mm (Materialprobe C) eingestellt.

4.2 Rheologische Methoden

Das EAM wurde durch oszillatorische Scherung simuliert. Die relativen Verschiebungen wurden während der Oszillation durch den Oszillationswinkel φ gesteuert. Für sämtliche Proben wurde das in Bild 7 dargestellte Messprofil angewandt und lediglich der Oszillationswinkel φ variiert. Für jede Materialprobe wurde ein Oszillationswinkel φ von 10° angewendet. Die Materialprobe B wurde zusätzlich einem Oszillationswinkel φ von 1,5° ausgesetzt. Dabei führte ein Oszillationswinkel φ von 1,5° zu einer Verschiebung zwischen der Materialprobe und dem äußeren Bereich des Messsensors von 4 mm (in Abhängigkeit vom Durchmesser des Probegefäßes und der Geometrie des Paddels), was eine mögliche Erweiterung des normativen Grenzwerts von 1 mm darstellt. Darüber hinaus wurde die Materialprobe B während der wiederkehrenden Oszillationsperiode einem Oszillationswinkel φ von 0° ausgesetzt, d.h. die Probe war während dieses Messabschnitts in Ruhe, um das ungestörte Materialverhalten zu erfassen. Die Frequenz f wurde für jede Messung zu 0,5 Hz gewählt. Das konzeptionelle Vorgehen des Untersuchungsprogramms ist dem Bild 6 zu entnehmen. Die rheologischen Messungen wurden mit einem Rotationsviskosimeter (Viskomat XL; Schleibinger Geräte Teubert und Greim GmbH) durchgeführt, das mit einem sogenannten Betonpaddel für Langzeitmessungen bestückt wurde. Aufgrund des Fehlens einer definierten Scherfläche und Spaltbreite können bei diesem Gerät relative Kenngrößen bestimmt werden. Das Drehmoment T (entsprechend der Scherspannung τ) wird gemessen, während die Umdrehungsgeschwindigkeit Ω (entsprechend der Scherrate $\dot{\gamma}$) des Probengefäßes gesteuert wird. Für jede Probe wurde die Messung 30 Minuten nach Wasserzugabe gestartet.

Während der Messung wurde die Temperatur der Proben mit einem Thermostat bei 20 °C konstant gehalten. Zu definierten Messzeitpunkten (vgl. Bild 7, oben) wurden das Fließdrehmoment T_0 und die Steigung der durch Rotation generierten Fließkurven bestimmt. Die Fließkurven wurden mit einem abwärts gerichteten Stufenprofil erzeugt (vgl. Bild 7, unten).

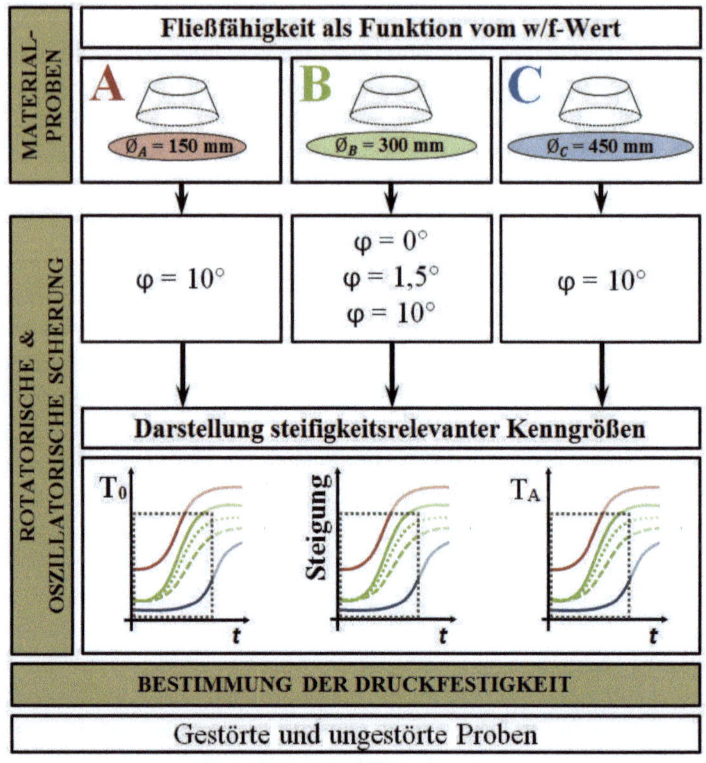

Bild 6: Konzeptionelles Untersuchungsprogramm. [5]

Zur Bestimmung des Fließdrehmoments T_0 wurde das HERSCHEL-BULKLEY-Modell angewendet, da dadurch das Fließdrehmoment T_0 des zum Teil scherverdickenden Fließverhaltens des Grout-Materials genauer erfasst werden konnte. Zur Abbildung der Viskosität wurde das BINGHAM-Modell über die Bestimmung der Fließkurvensteigung ermittelt. Unter Anwendung des HERSCHEL-BULKLEY-Modells waren unerwartet hohe Streuungen der plastischen Viskosität μ beobachtet worden.

Jeder rotatorischen Messung (Messdauer von 4 min) folgte eine Oszillationsperiode einer Messdauer von 20 min. Rotation und Oszillation wechselten sich bis zum Ende der Messung ab. Um die Steifigkeitszunahme des Materials im Zeitverlauf darzustellen, wurden das Fließdrehmoment T_0, die Steigung der Fließkurven ($\Delta\Omega/\Delta T$) und die Drehmomentamplitude T_A aus der Oszillationsperiode als Funktion von der Zeit erfasst (vgl. Bild 6). Um den Einfluss des EAM auf die mechanischen Eigenschaften erfassen zu können, wurden die Materialproben nach der Messung (gestörte Proben) verwendet, um Probekörper zur Ermittlung der Druckfestigkeit herzustellen. Zusätzlich wurden Probekörper aus ungestörtem Material jeder Materialprobe als Referenz hergestellt. Von jeder Materialprobe (gestört und ungestört) wurden drei Probekörper mit einer Kantenlänge von 40 mm x 40 mm x 160 mm hergestellt und nach 7 Tagen gemäß DIN EN 196-1 geprüft. Die Proben wurden bis zur Prüfung unter Wasser gelagert.

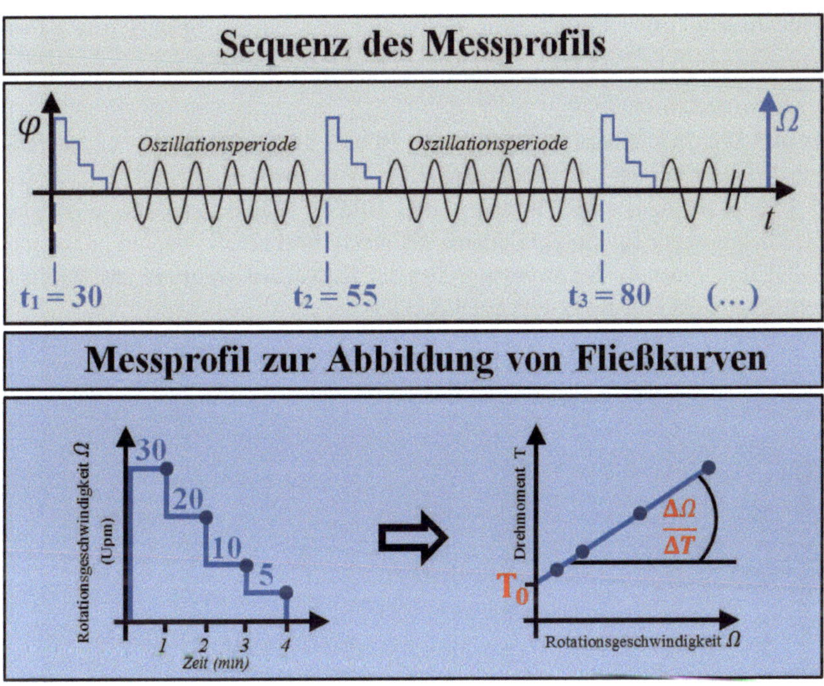

Bild 7: Messprofile der oszillatorischen und rotatorischen Scherversuche. [5]

5 Versuchsergebnisse und Diskussion

Die folgende Tabelle zeigt den w/f-Wert zur Einstellung der Fließfähigkeit des verwendeten Materials. Sämtliche Materialproben zeigten ein entmischungsfreies Fließen.

Tabelle 1: Resultierende Fließfähigkeiten der Materialproben in Abhängigkeit vom Wassergehalt. [5]

Wassergehalt	24,0 %	27,4 %	30,5 %
Ausbreitfließmaß	150 mm	300 mm	450 mm

Der w/f-Wert beeinflusst maßgeblich die Rheologie des Materials. Mit steigendem w/f-Wert nimmt das Fließdrehmoment T_0 ab, was zu einem Rückgang der Fließfähigkeit führt (vgl. Tabelle 1 und Bild 8). Bild 8 zeigt das ermittelte Fließdrehmoment T_0 im Zeitverlauf. Es kann beobachtet werden, dass das Fließdrehmoment T_0 der Materialproben im Zeitverlauf aufgrund des Hydratisierungsprozesses überproportional ansteigt.

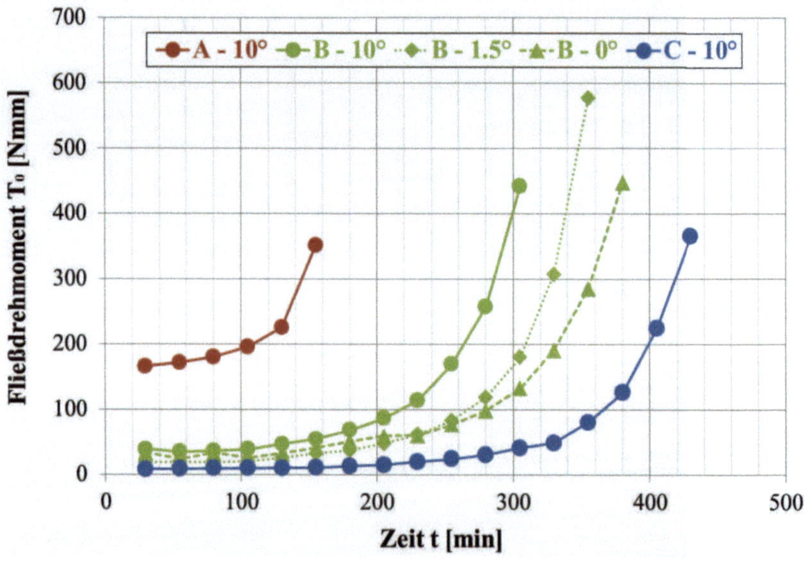

Bild 8: Entwicklung des Fließdrehmoments T_0 als Funktion von der Zeit. [5]

Das Fließdrehmoment T_0 der Materialprobe A nimmt bis zum Erreichen der maschinenspezifischen Drehmomentbegrenzung in einem Zeitraum von 155 min ca. um den Faktor 2 zu. Das Fließdrehmoment T_0 der Materialproben B (0°, 1,5°, 10°) nimmt in einem Zeitraum zwischen 305 min und 380 min um das ca. 11- bis 30-fache zu. Aufgrund der geringen Anfangssteifigkeit der Materialprobe C kann eine Zunahme des Fließdrehmoments T_0 um das 46-fache beobachtet werden, wobei das Fließdrehmoment T_0 über einen Zeitraum von 430 min erfasst wurde.

Wie zu erwarten wird mit kleinerem w/f-Wert ein höheres anfängliches Fließdrehmoment T_0 gemessen und umso früher beginnt die Zunahme des Fließdrehmoments T_0. Bei einem niedrigeren w/f-Wert ist der Abstand zwischen den Zementpartikeln in der Suspension des Grout-Materials geringer. Infolgedessen können Porenräume durch das Wachstum früher Hydratationsprodukte schneller überbrückt werden, was zur Agglomeration von Partikeln führt. Dieser Vorgang erhöht das Fließdrehmoment T_0 der Probe zu einem früheren Zeitpunkt.

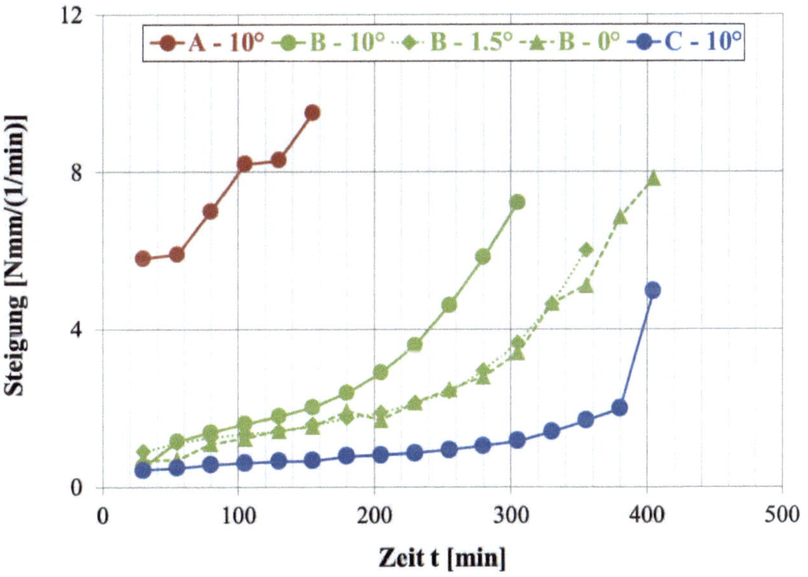

Bild 9: Entwicklung der Viskosität (repräsentiert als Steigung der Fließkurven) als Funktion von der Zeit. [5]

Außerdem ist zu erkennen, dass ein größerer Oszillationswinkel φ zu einer beschleunigten Zunahme der rheologischen Eigenschaften (Fließdrehmoment T_0 und Viskosität (als Steigung der Fließkurven)) führt (vgl. Bild 8 und Bild 9). Dieses Verhalten kann durch zwei Phänomene erklärt werden. Zum einen werden Primärstrukturen durch die von außen aufgebrachte Scherung aufgrund einer Erhöhung der Impulsenergie gebildet. Je größer der Oszillationswinkel φ ist, desto größer ist der Energieeintrag in das System, was zu einer erhöhten Agglomeration von Partikeln führt. Andererseits führt die Scherbeanspruchung zu einem Entlüftungsprozess (vor allem bei sehr steifen Proben), der jedoch auf die frühe Belastungsphase beschränkt ist. Dies bestätigt die Beobachtungen von [19] und [20] (vgl. Bild 2). Luftblasen im Material können unter Scherung als Schmierung wirken. Kommt es zur Entlüftung des Materials, erhöht sich die Steifigkeit und dadurch das gemessene Drehmoment. Das Phänomen der Entlüftung spiegelt sich auch in den Ergebnissen der Druckfestigkeit wider, auf die später näher eingegangen wird.

In Anbetracht beider Effekte wird geschlussfolgert, dass insbesondere die durch Oszillation hervorgerufene Agglomeration für den starken Anstieg des Fließdrehmoments T_0 im höheren Probenalter relevant ist. Neben der Erhöhung des Fließdrehmoments T_0 kann eine Erhöhung der Viskosität beobachtet werden, worauf die Änderung der Steigung der Fließkurven schließen lässt. Der Einfluss des Oszillationswinkels φ wird durch die Ergebnisse in Bild 10 bestätigt. Es kann auch bestätigt werden, dass mit zunehmendem w/f-Wert eine beschleunigte Steifigkeitszunahme des Materials einhergeht, was sich in der Steigung der Fließkurven widerspiegelt. Je geringer der Partikelabstand, desto früher kann eine Steifigkeitszunahme beobachtet werden.

Während die Ergebnisse in Bild 8 und Bild 9 durch rotationsgesteuerte Messungen nach definierten Oszillationsphasen ermittelt wurden, zeigt Bild 10 die Ergebnisse der Oszillationsversuche.

Dargestellt ist die zeitliche Änderung der Drehmomentamplitude T_A bei definiertem Oszillationswinkel φ. Die Drehmomentamplitude T_A wurde alle 5 Minuten bestimmt. Im Hinblick auf die Ergebnisse der Oszillationsmessungen können die vorgenannten Zusammenhänge bestätigt werden. Es ist zu beobachten, dass die als Drehmomentamplitude T_A dargestellte Steifigkeit sämtlicher Proben während der Oszillationsperiode mit der Zeit ansteigt.

Weiterhin kommt es als Folge der eingeschobenen Rotationsperiode in der darauffolgenden Oszillationsperiode zu einer kurzzeitigen Abnahme der Drehmomentamplitude T_A (Strukturbruch) und das sogar bei fortgeschrittenem Hydratationsgrad, also bei hoher Materialsteifigkeit. Infolge eines erhöhten Energieeintrags kommt es also trotz hoher Steifigkeit zu einer kurzzeitigen Verflüssigung des Materials.

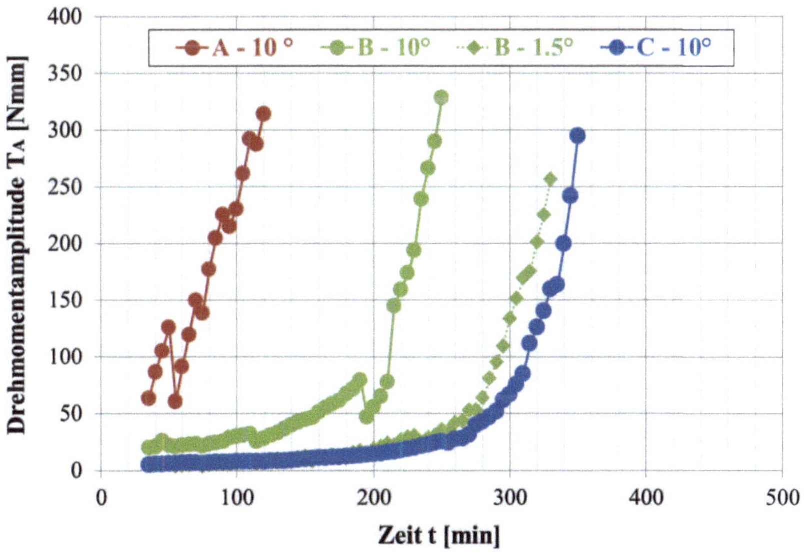

Bild 10: Entwicklung der Drehmomentamplitude T_A aus Oszillationsversuchen als Funktion von der Zeit. [5]

Bild 11 zeigt die gemessenen Druckfestigkeiten der nach den rheologischen Messungen hergestellten und über einen Zeitraum von 7 Tagen erhärteten Probekörper. Wie zu erwarten ist der Einfluss des w/f-Wertes auf die Druckfestigkeit zu erkennen. Je höher der Wassergehalt, desto geringer ist die Druckfestigkeit des Materials. Die gemessenen Druckfestigkeiten bestätigen darüber hinaus das Phänomen der Entlüftung während der Oszillation, auch wenn kein direkter Beweis über die Bestimmung des Luftporengehalts geliefert wird. Es kann beobachtet werden, dass die Proben, die durch die oszillatorische Scherung kontinuierlich gestört wurden, ausnahmslos eine höhere Druckfestigkeit aufweisen als die ungestörten Proben.

Dies lässt den Schluss zu, dass die kontinuierliche, zyklische Belastung einen Prozess der Entlüftung des Grout-Materials verursacht. Außerdem kann beobachtet werden, dass der Einfluss von zyklischen Bewegungen auf den Entlüftungsprozess umso geringer ist, je höher die anfängliche Fließfähigkeit des Materials ist.

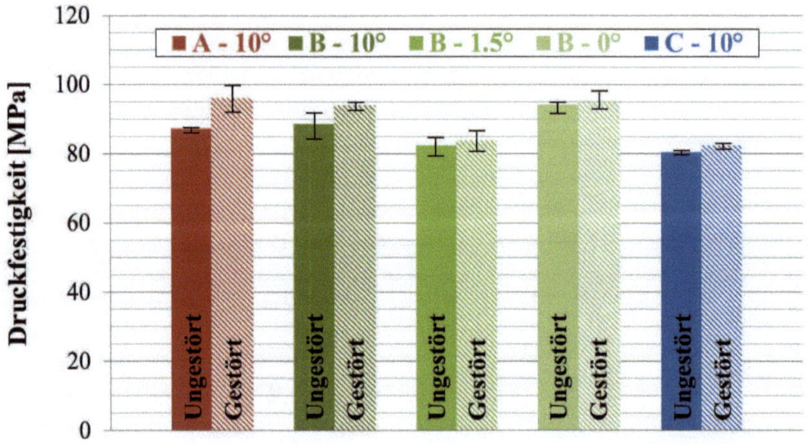

Bild 11: Druckfestigkeit der durch Oszillation gestörten und ungestörten Proben. [5]

Dementsprechend zeigt die Materialprobe C einen geringeren Unterschied in der gemessenen Druckfestigkeit (ungestört und gestört), da die hohe Fließfähigkeit bereits zu einem ausreichend starken Entlüftungsprozess der ungestörten Proben führt. Es kann somit festgestellt werden, dass bei keinem der Versuche eine Abnahme der Druckfestigkeit in Folge der strukturstörenden Scherung beobachtet werden konnte.

6 Zusammenfassung und Schlussfolgerungen

Der Einfluss des EAM auf die Tragfähigkeit von Offshore-Tragstrukturen ist bisher kaum erforscht. Aus diesem Grund legen einschlägige Normen fest, dass während der ersten 24 Stunden der Installationsphase die maximale Relativbewegung zwischen den Stahlrohrelementen einer Grout-Verbindung auf 1 mm zu begrenzen ist. Diese Anforderung führt in der Planungs- und Ausführungsphase zu erheblichen Einschränkungen.

Die maximale Relativverschiebung ist bislang unter Berücksichtigung der zu erwartenden maximalen Wellenhöhen für eine unvergossene Grout-Verbindung unter Anwendung numerischer Simulationen zu berechnen. Es ist jedoch offensichtlich, dass das in den Ringspalt eingefüllte Grout-Material zu einer erheblichen Erhöhung der Steifigkeit der Struktur führt und somit die maximal zu erwartende Relativverschiebung reduziert wird. Ansätze zur Berücksichtigung der Steifigkeit des Grout-Materials werden derzeit jedoch nicht von der normativen Seite bereitgestellt. In diesem Artikel wurde eine Methode zur Berücksichtigung des Phänomens EAM und gleichzeitig zur Bestimmung von steifigkeitsrelevanten Materialkennwerten vorgestellt. Um das EAM während des Ansteifens zu simulieren, wurden oszillatorische Scherversuche durchgeführt. Zur Bestimmung von steifigkeitsrelevanten Materialkennwerten wurden sowohl rotations- als auch oszillationsgesteuerte Scherversuche angewandt. Für die Untersuchungen wurden variierende Fließfähigkeiten durch Anpassung des w/f-Wertes und unterschiedliche Oszillationswinkel φ berücksichtigt.

Es hat sich gezeigt, dass durch den w/f-Wert nicht ausschließlich die Fließfähigkeit und somit die anfängliche Steifigkeit des Grout-Materials gesteuert werden kann, sondern auch Einfluss auf die Steifigkeitsentwicklung genommen werden kann. Wie zu erwarten, ist die anfängliche Materialsteifigkeit umso größer, je niedriger das Verhältnis von Wasser zu Feststoff ist. Darüber hinaus führt ein niedrigerer w/f-Wert zu einer beschleunigten Erhöhung der Materialsteifigkeit. Eine wesentliche Erkenntnis dieser Untersuchung ist, dass ein größerer Oszillationswinkel φ, der zu einer größeren Verschiebung führt, eine beschleunigte Zunahme der Materialsteifigkeit zur Folge hat, was auf eine Entlüftung des Materials (hauptsächlich in der frühen Phase der Belastung) zurückzuführen ist. Überlagert wird dieser Prozess durch extrinsische Agglomerationsprozesse. Insbesondere der Entlüftungsprozess führt zu einer Erhöhung der Festigkeit des ausgehärteten Materials besonders bei niedrigen w/f-Werten.

Grundsätzlich zeigen die Ergebnisse, dass die Methode zur Nachstellung des EAM mittels oszillatorische Scherversuche vielversprechend ist. Auch wenn die Oszillation die Offshore-Realität nicht vollständig abbildet, kommt diese Art von zyklischer Bewegung den wellenbedingten Belastungen sehr nahe. Zusätzlich können auf diese Weise variierende Oszillationswinkel φ und Frequenzen abhängig vom zu erwartenden Seegang simuliert werden. Die Kombination aus Rotation und Oszillation ermöglicht ein tiefes Verständnis über das rheologische Verhalten des Grout-Materials.

Auch wenn die gesamte Zeitspanne von 24 Stunden (gemäß DNV-OS-J101 [8]) mit der vorgestellten Methodik nicht simuliert und gemessen werden kann, zeigen die Versuchsergebnisse, dass mit solchen Versuchen der Einfluss des EAM quantifiziert werden kann.

Als Empfehlung für die Praxis kann festgehalten werden, dass mit geringerem w/f-Wert (also mit geringerer Anfangsfließfähigkeit) Entmischungserscheinungen vorgebeugt werden kann, dass die Steifigkeitszunahme beschleunigter abläuft und dass die Druckfestigkeit des ausgehärteten Materials höher ist. Natürlich muss die Fließfähigkeit des Grout-Materials hoch genug sein, damit es ohne Störungen gepumpt werden kann.

7 Ausblick

Weitere Untersuchungen sind geplant, um auch das Materialverhalten über die hier betrachtete Zeitspanne hinaus zu erfassen, indem die Methodik durch Ultraschallmessungen erweitert wird. Es ist beabsichtigt das EAM auch während der Ultraschallmessungen realitätsnah zu simulieren. Als Bindeglied zwischen rheologischem Verhalten und der Druckfestigkeit sollen Ultraschallmessungen durchgeführt werden.

Dadurch soll ein umfassendes Eigenschaftsprofil des Grout-Materials vom flüssigen bis zum erhärteten Zustand abgebildet werden. Sämtliche Materialkennwerte sollen anschließend als Eingangsgrößen für numerische Simulationen verwendet werden, um die am Bauwerk maximal zu erwartende Relativverschiebung auf realistische Weise vorherzusagen.

8 Danksagung

Die Untersuchungen wurden im Forschungsprojekt GREAM durchgeführt. Das Forschungsprojekt (Förderkennzeichen: 0324257) wird vom Bundesministerium für Wirtschaft und Energie (BMWi) gefördert und vom Projektmanagement Jülich (PtJ) koordiniert. Das Forschungsprojekt wird in Kooperation mit dem Institut für Stahlbau der Leibniz Universität Hannover durchgeführt.

Ein besonderer Dank gilt dem Materialhersteller für die Bereitstellung der verwendeten Materialien.

Literatur

[1] Al-Khazali, H.A.H.; Askari, M.R.: Geometrical and Graphical Representations Analysis of Lissajous Figures in Rotor Dynamic System, IOSR Journal of Eng, 2(5), 971-978, 2012.

[2] Banfill, P.F.G.: Rheology of fresh cement and concrete, The British Society of Rheology, London, 2006.

[3] Billington, C.J.; Tebbett, I.E.: The Basis for New Design Formulae for Grouted Jacket to Pile Connections, Proc 12th Annual Offshore Technology Conf, Houston, Texas, USA, 1980.

[4] Cotardo, D.; Lohaus, L.: Zusammenhang von plastischer Viskosität, Strukturabbaurate und Mischungsstabilität von Bindemittelsuspensionen, Proc 25th Conf and Laboratory Workshops, Regensburg, Germany, 120-138, 2016.

[5] Cotardo, D.; Haist, M.; Lohaus, L.: Early-age Movement in Grouted Joints for Offshore Applications – Determiation of the Development of Grout-stiffness, Proc 29th Int Offshore and Polar Eng Conf, ISOPE, Honolulu, Hawaii, USA, 2019.

[6] de Larrard, F.; Ferraris, C.F.; Sedran, T.: Fresh concrete: A Herschel-Bulkley material, Material and Structures, 31(7), 494-498, 1998.

[7] Derjaguin, B.V.: Friction and adhesion, Iv. The theory of adhesion of small particles, Kolloid Zeits, 69, 155-164, 1934.

[8] Det Norske Veritas: DNV-OS-J101 – Design of Offshore Wind Turbine Structures, 2014.

[9] DIN EN ISO 19902:2014-01: Petroleum and natural gas industries - Fixed steel offshore structures, 2014.

[10] DIN EN 196-1:2016-11: Methods of testing cement - Part 1: Determination of strength, 2016.

[11] DIN EN 1015-3:2007-05: Methods of test for mortar for masonry - Part 3: Determination of consistence of fresh mortar (by flow table), 2007.

[12] Haist, M.: Zur Rheologie und den physikalischen Wechselwirkungen bei Zementsuspensionen, Karlsruher Reihe Massivbau, Baustofftechnologie, Materialprüfung, 66, KIT Scientific Publishing, Karlsruhe, 2010.

[13] Ingebrigtsen, T.; Løset, Ø.; Nielsen, S.G.: Fatigue Safety and Overall Safety of Grouted Pile Sleeve Connections, Proc 22nd Annual Offshore Technology Conf OTC, OTC paper 6344, Houston, Texas, USA,1990.

[14] Kallmann, H.; Willstaetter M.: Zur Theorie des Aufbaus kolloidaler Systeme, Die Naturwissenschaften, 16(51) 952-953, 1932.

[15] Lamport, W.B.; Jirsa, J.O.; Yura, J.A.: Grouted Pile-to-Sleeve Connection Tests, Report on a Research Project, PMFSEL Report, 86(7), Department of Civil Engineering, University of Austin in Texas, United States of America, 1986

[16] Lee, H.K.; Lee, K.M.; Kim, Y.H.; Yim, H.; Bae, D.B.: Ultrasonic in-situ monitoring of setting process of high-performance concrete, Cem. Concr. Res., 34(4), 631-640, 2004.

[17] Lohaus, L.; Cotardo, D.; Werner, M.: The Early Age Cycling and its Influence on the Properties of hardened Grout Material, Proc Int 1st Wind Eng Conf, IWEC, Hanover, Germany, 2014.

[18] Lohaus, L.; Cotardo, D.; Werner, M.: A test system to simulate the influence of early age cycling on the properties of grout materials, Proc 24th Int Offshore and Polar Eng Conf, ISOPE, Busan, 4, 234-239, 2014.

[19] Lohaus, L.; Schaumann, P.; Cotardo, D.; Kelma, S.; Werner, M.: Experimental and Numerical Investigations on Grouted Joints in Monopiles Subjected to Early-age Cycling to Evaluate the Influence of Different Wave Loadings, Proc 25th Int Offshore and Polar Eng Conf, ISOPE, Kona, Hawai'i, USA, 1, 268-276, 2015.

[20] Lohaus, L.; Cotardo, D.; Werner, M.; Schaumann, P.; Kelma, S.: Experimental and Numerical Investigations of Grouted Joints in Monopiles Subjected to Early-Age Cycling, Int J Offshore and Polar Eng, JOWE, 2(4), 193-201, 2015.

[21] Lowke, D.: Sedimentationsverhalten und Robustheit Selbstverdichtender Betone – Optimierung auf Basis der Modellierung der interpartikulären Wechselwirkungen in zementbasierten Suspensionen, Dissertation, München, 2013.

[22] Mewis, J.: Thixotropy – General-Review, J. Non-Newtonian Fluid Mech; 6, 1-20,1979.

[23] Mezger, T.G.: The Rheology Handbook: For users of rotational and oscillatory rheometers, Vincentz Network, 2002.

[24] Min, B.-H.; Erwin, L.; Jennings, H.M.: Hysteresis loops of cement paste measured by oscillatory shear experiments, Korean Journal of Rheology, 5(2), 99-108, 1993.

[25] Powers, T.C.: The properties of fresh concrete, John Wiley & Sons, New York, 1986.

[26] Reinhardt, H.W.; Grosse, C.U.: Continuous monitoring of setting and hardening of mortar and concrete, Constr. Build. Mater., 18(3), 145-154, 2004.

[27] Robeyst, N.; Gruyaert, E.; Grosse, C.U.; De Belie, N.: Monitoring the setting of concrete containing blast-furnace slag by measuring the ultrasonic p-wave velocity, Cem. Concr. Res., 38(10), 1169-1176, 2008.

[28] Schultz, M.A.; Struble, L.J.: Use of oscillatory shear to study flow behavior of fresh cement paste, Cem. Concr. Res., 23(2), 273-282, 1992.

[29] Tattersall, G.H.; Banfill, P.F.G.: The Rheology of Fresh Concrete, Pitman Books Limited, Great Britain, 1983.

[30] Verwey, E.J.W.: Theory of stability of lyophobic colloids, Amsterdam, 1984.

[31] Wallevik, J.E.: Rheological properties of cement paste: Thixotropic behavior and structural breakdown, Cem. Concr. Res., 39, 14-29, 2009.

Rheology as a Tool to Characterize Dissolution and Reappearance of Air under Pressure in Cement Pastes

Daniel Galvez Moreno[1)], Kyle Riding[2)], Dimitri Feys[1)]
[1)] Department of Civil, Architectural and Environmental Engineering, Missouri University of Science and Technology, Rolla, MO, United States
[2)] Department of Civil and Coastal Engineering, University of Florida, Gainesville, FL, United States
Phone: +1 573 341 7193, e-mail: feysd@mst.edu

Abstract

Pumping is one of the most efficient ways to place concrete in a formwork. Due to pressure and high shear rates during pumping processes, some concrete properties change. One of them is the air-void system, which is reported to coarsen after pumping, reducing the freeze-thaw resistance of concrete. This paper utilizes rheology as a tool to investigate bubble dissolution and reappearance under pressure, by taking the concepts of Henry's law and the capillary number into consideration. For the evaluated cement pastes, viscosity and relative shear stress decrease with applied pressure, indicating a portion of the air bubbles dissolve. Releasing the pressure causes a (partial) regain of the viscosity, due to reappearance of air.

1 Introduction

In countries with cold climates experiencing numerous freeze-thaw cycles, such as Canada, the United States and Northern Europe, air-entraining agents are frequently used to prevent freeze-thaw and scaling damage in concrete [1]. A good air-void system in concrete limits the distance freezing water needs to travel in concrete pores to reach the nearest air bubble, relieving pressure and tensile stresses executed on the pore walls [1]. The spacing factor is an average measure of this travel distance, and some specifications and recommendations list 200 or 230 μm as an upper limit. These low spacing factors are typically achieved when the air in concrete shows a distribution of many small air bubbles. However, most specifications impose a target air content for concrete subject to freeze-thaw cycles [2]. Furthermore, this is measured on a sample of fresh concrete before it is placed, consolidated and finished. As such, if the air-void system can change during these latter processes, there is no guarantee that the air-void system is adequate in the concrete structure, even if the air content test yielded an acceptable result.

Pumping of concrete is a process which is known to influence the air-void system [3-6]. However, an accurate prediction tool or equation is not available at this time. In a majority of cases, the spacing factor increases while the air content decreases, but it is unknown in which situations this exactly happens, and for which parameters the magnitude of these changes is unacceptable. Recent research work has been performed based on the findings and recommendations obtained over two decades ago [7]. This latter work shows that the evaluated self-consolidating concrete (SCC) mixtures show a much higher sensitivity to a change in spacing factor compared to the tested conventional vibrated concrete mixtures (CVC) [7]. However, further work is necessary to discover the mechanisms causing these changes.

This paper shows the first steps in the development of a theoretical framework to predict changes in the air-void system due to pumping operations, by means of rheological measurements on cement pastes, executed at atmospheric and elevated pressure. After the introduction of the theoretical concepts governing the behavior, the testing methodology and results are discussed.

2 Theoretical Concepts

Three theoretical concepts are introduced in this section to explain the rheological behavior of cement pastes under pressure: the universal gas law, Henry's law and the behavior of gas bubbles under shear.

2.1 Universal Gas Law

The universal gas law states that the product of pressure and volume of a gas is equal to the number of moles of gas, multiplied with Boltzmann's constant and temperature. For concrete pumping operations, the mass of gas remains constant in the pipes (conservation of mass), and temperature is approximately constant. This means that the volume of gas is inversely proportional to the applied pressure. As such, to reduce all air bubbles' radii by a factor 2, pressure needs to be increased by a factor 8. However, in case of under-pressure (suction) bubbles will grow and can potentially coalesce.

2.2 Henry's Law

Henry's law states that, in steady-state conditions and at constant temperature, the partial pressure of a gas is proportional to Henry's constant and the concentration of dissolved gas [3, 8]. In other words, the maximum concentration of dissolved gas a liquid can carry is linearly proportional to the applied pressure (assuming the bubbles are not too small). Increasing pressure will lead to a visible reduction of air molecules in the gas phase.

2.3 Bubbles under Shear

Air-entrainment is generally accepted as a method to reduce the viscosity of concrete. However, this is not universally true! The viscosity reduction is the consequence of air bubbles deforming and aligning with the flow direction. At rest, air bubbles are preferentially in a spherical form. This is caused by the surface tension between the air and the surrounding liquid. If the shear stresses are low compared to surface tension, the air bubble will remain spherical, and the viscosity reduction will be transformed in a viscosity increase [9]. This can be explained as follows: If the bubbles are non-deformable, they act as solid particles. An increase in solid particles in a suspension causes an increase in viscosity [10]. To predict the deformability behavior of air in liquids, one can calculate the Capillary number (Ca):

$$Ca = \frac{\dot{\gamma}\eta a}{\Gamma} \tag{1}$$

Where Ca is the capillary number (-), $\dot{\gamma}$ is the shear rate in s^{-1}, η is the apparent viscosity of the suspending medium (Pa s), a is the bubble diameter (m) and Γ is the surface tension at the water-air interface (N/m). If Ca is much larger than 1, the bubbles deform [9, 11]. In the opposite case of Ca much smaller than 1, the bubbles remain spherical [12].

3 Rheological Measurements

The rheological measurements were performed with an Anton Paar MCR 302, equipped with a pressure cell. The configuration inside the pressure cell was a concentric cylinder system with a gap size of approximately 1 mm. The imposed procedure was a constant shear rate of 100 s^{-1}, with three intermediate flow curves, with a decreasing shear rate from 100 to 0.1 s^{-1}: one before, one during and one after pressurizing the sample. The differential viscosity was calculated by differentiating the best fitting 6th order polynomial to the shear rate and evaluating this function at a shear rate of 50 s^{-1}. The pressure was imposed stepwise, by manually adjusting the pressure regulator. Pressure increments were in the order of 2-4 bar, with a maximum around 10 bar, which is mathematically sufficient to dissolve all air in the water of the evaluated pastes. At depressurization, all pressure is evacuated at once. All measurements were performed at constant temperature of 20 °C.

4 Cement Pastes

Two series of cement pastes were produced, one at w/c = 0.45 and one at w/c = 0.35. The cement was a commercially available ordinary Portland cement and no supplementary cementitious materials or mineral fillers were employed. Air-entraining agent (AEA) dosages were varied to obtain different air contents, while the dosage of dispersing admixture (SP) was adjusted to obtain a constant mini-slump flow, corresponding to a relatively flowable mixture to guarantee adequate rheological measurements. A retarder was added to minimize the effect of hydration on the overall experiment. Table 1 shows the mix designs of the two cement pastes, with the variations in air content and AEA dosage.

Table 1: Mix design of cement pastes. All units are in g/l, unless expressed otherwise. AC stands for air content.

	Paste w/c = 0.45	**Paste w/c = 0.35**
Cement	1264.5	1452.9
Water	569.0	508.5
SP	1.644	3.196
Retarder	5.057	5.812
AEA – AC - 1	2.023 / 4.4%	11.624 / 4.4%
AEA – AC - 2	6.068 / 5.6%	11.624 / 6.3%
AEA – AC - 3	10.112 / 7.1%	23.233 / 7.2%

All mixing was performed in a small Hobart mixer, by adding the water and AEA to the bowl first, followed by all cement, and mixing at speed 1 (134 rpm). After scraping the bowl for 30 s, the SP was added and mixing resumed for 2 minutes. Halfway through the second mixing step, the retarder was added without interrupting the mixing. The mixing speed was speed 1 for all w/c = 0.45 mixtures and the first w/c = 0.35 mixture. For the remaining 0.35 mixtures, the speed was increased to 285 rpm to entrain more air, hence explaining the difference between the first and second 0.35-w/c-mixtures having identical AEA dosages. Air content was determined according to ASTM C185, without imposing any consolidation.

5 Results

5.1 Effect of Pressure on Shear Stress Evolution with Time

Figure 1 shows a typical evolution of the shear stress with time, including the determination of the three flow curves for the third mixture with w/c = 0.45. The preshear period was extensive to ensure equilibrium of the sample. The stress shown is the relative stress, determined by dividing the shear stress at time t by average of the shear stress measured five to zero seconds before the first application of pressure. Two curves on two samples of the same cement paste are shown: one which undergoes the pressure steps while the other shows the behavior of a non-pressurized sample.

From Figure 1, it can clearly be seen that with the application of pressure, the relative shear stress decreases gradually with time, and does so with the application of the second and third (although very slightly) pressure step. The reason why a decrease in relative shear stress is observed is because the Ca-number of a 500 µm bubble in that mixture equals 0.19, indicating most air bubbles remain spherical during flow. Removing air bubbles results thus in a decrease in viscosity, which is reflected in the decrease in relative stress at constant shear rate. Any further increase beyond 8 bar causes no additional change in relative shear stress, which can be attributed to all air being dissolved. Releasing the pressure causes the relative shear stress to increase back to the original level, or close to that level for the 2nd and 3rd mixtures with w/c = 0.35. This increase in shear stress is associated with the reappearance of the air. However, with further elapsed time, not shown in the graph, a more complicated pattern of decreasing and increasing shear stress is observed, which is assumed to be caused by escaping air bubbles (as they may no longer be stabilized by the AEA), and segregation of the cement particles after that.

Figure 2 shows the relative shear stress of all pressurized samples during the first pressurizing steps: the mixtures with w/c = 0.35 on top, those with w/c = 0.45 on the bottom graph. The graphs indicate a decrease of shear stress under pressure for all pastes. However, for the pastes with w/c = 0.35 and air contents above 6%, the decrease in shear stress is no longer gradual. It is hypothesized that due to the finer air void distribution in those latter pastes, the small bubbles dissolve practically immediately. Indeed, the % of air bubbles smaller than 500 μm, measured by the fresh concrete air-void analyzer, are 74, 72, and 70% for the three mixtures with w/c = 0.45, respectively, and 57, 79 and 89% for the mixtures with w/c = 0.35, respectively. Further research is needed to confirm this hypothesis.

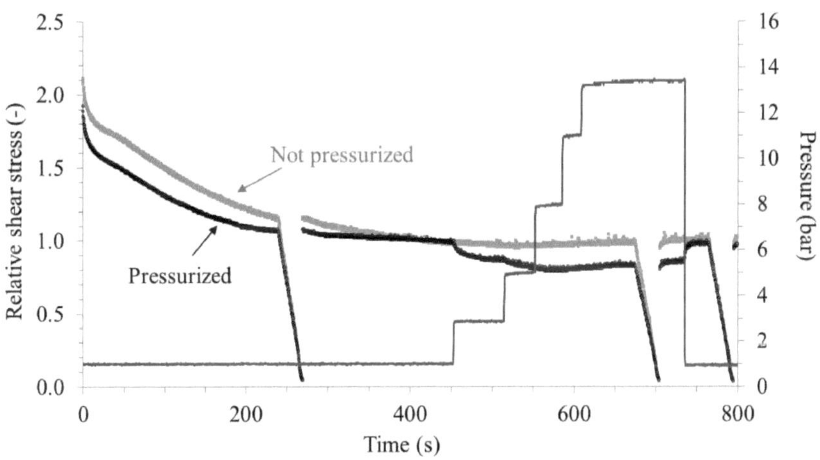

Figure. 1: Relative shear stress for the pressurized (black) and non-pressurized (grey) sample over time. The stepwise curve, starting around 420 s, reflects the applied pressure.

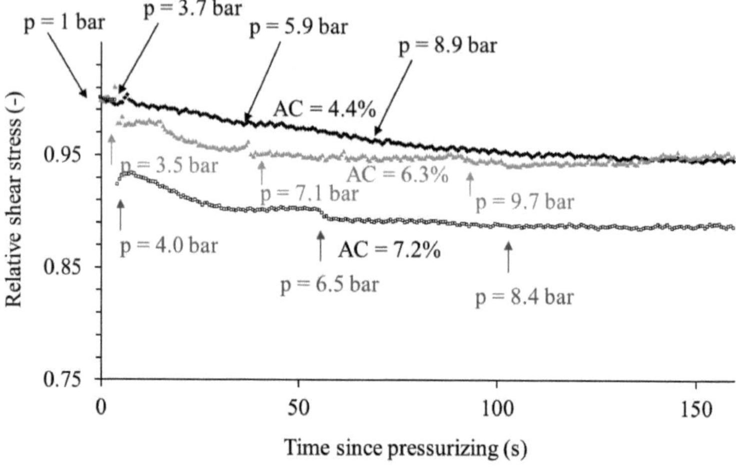

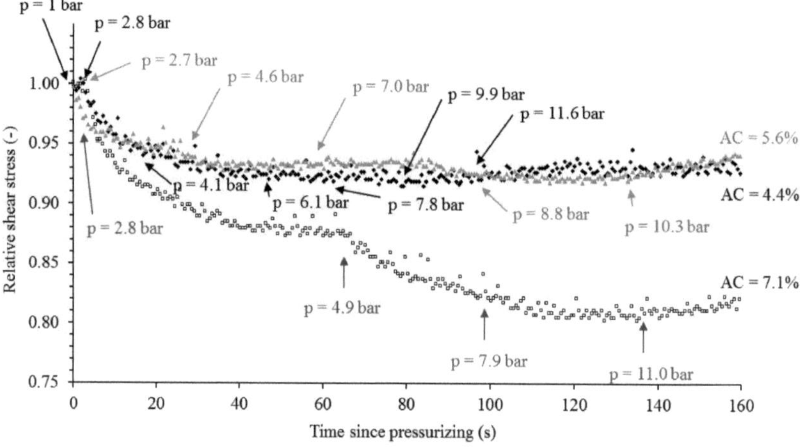

Figure. 2: Relative shear stress over time since pressurization, for the mixtures with w/c = 0.35 (top) and w/c = 0.45 (bottom). The arrows indicate when a higher pressure was imposed for each mixture.

5.2 Effect of Pressure on Differential Viscosity

All six mixtures were subjected to three flow curve measurements, one before, one during and one after the pressurization. The flow curve during pressurization was determined at the highest pressure for that each test (10 ± 2 bar). Table 2 shows the results of the differential viscosities calculated at 50 s^{-1}. For each mixture, it can be seen that the viscosity during pressurization is lower than the viscosity before the application of pressure, reflecting the relative shear stress results. As the non-pressurized samples (not shown) do not show such change, this effect can solely be attributed to the compression and dissolution of air in the sample. When removing the pressure, the w/c = 0.45 samples show a regain of viscosity, close to the viscosity before the pressure step. However, only the sample at the highest air content for the w/c = 0.35 mixtures shows some regain in viscosity. The lack of full regaining of the viscosity may be attributed to the escape of air bubbles prior to determining the last flow curve. When opening the pressure cell after the experiment, a foam is observed on top of the inner cylinder (Figure 3), indicating that air has escaped during the experiment, most likely just after depressurization.

Table 2: Differential viscosity for all mixtures before, during and after the pressurization steps.

		Differential Viscosity (Pa s)		
		Before pressure	During pressure	After pressure
w/c = 0.45	AC = 4.4%	0.163	0.144	0.156
	AC = 5.6%	0.176	0.158	0.171
	AC = 7.1%	0.159	0.122	0.150
w/c = 0.35	AC = 4.4%	0.558	0.488	0.490
	AC = 6.3%	0.499	0.482	0.477
	AC = 7.2%	0.509	0.447	0.474

6 Conclusions

For six different pastes, with three different air contents at two different w/c, the rheological properties were determined under pressure and compared to non-pressurized samples. Applying pressure to an air-entrained sample increases the capacity of the water in the paste to carry more dissolved air, according to Henry's law. Also, for the considered cement pastes, the capillary number is sufficiently small, indicating that the air bubbles remain spherical under the imposed shearing conditions.

The results from the differential viscosity measurements show that the viscosity has decreased when pressure is applied to the sample, which is in line with the reduction of "stiff air particles" in the mixture, according to the above-mentioned laws. After releasing the pressure, the viscosity is recovered, at least up to a certain extent.

Figure. 3: A substantial amount of foam is observed on the inner cylinder when opening the pressure cell after execution of the entire procedure.

The shear stress evolution with time, measured at constant shear rate of 100 s^{-1}, delivers similar conclusions: the shear stress decreases with time each time additional pressure is applied, up to a certain threshold pressure above which no further changes are observed. The decrease in stress is in agreement with the results on viscosity, and the threshold pressure could indicate that maximum dissolution is achieved. After releasing the pressure, the shear stress suddenly increases, which is associated with the reappearance of the air, followed by a decrease and increase in shear stress, which is assumed to be caused by rising of (non-stabilized) air bubbles, and segregation of cement particles, respectively.

One aspect for further research is the sudden decrease in shear stress with the first application of pressure for the mixtures with a finer air-void system, while for mixtures with a slightly coarser system, the decrease in shear stress is more gradual with time when pressure is suddenly applied.

7 Acknowledgments

The authors would like to acknowledge the Research Foundation of the American Concrete Institute and the RE-CAST Tier-1 University Transportation Center for the financial support, as well as the Center of Infrastructure Engineering Studies for the use of the facilities.

Bibliography

[1] Neville, A.M.: Properties of Concrete, Longman, London, 1995.

[2] ACI Committee 318: Building code requirements for structural concrete and commentary, ACI, Farmington Hills, 2014.

[3] Elkey, W., Janssen, D. J., Hover, K. C.: Concrete pumping effects on entrained air-voids. Washington State Department of Transportation, Washington State Department of Transportation, Washington State Transportation Commission, Transit, Research, and Intermodal Planning (TRIP) Division, 1994.

[4] Pleau, R., Pigeon, M., Lamontagne, A., Lessard, M.: Influence of pumping on characteristics of air-void system of high-performance concrete, Transportation research record, 30-36, 1995.

[5] Pleau, R., Pigeon, M.: Durability of concrete in cold climates, CRC Press, 2014.

[6] Hover, K. C., Phares, R. J.: Impact of concrete placing method on air content, air-void system parameters, and freeze-thaw durability, Transportation Research Record, 1532(1), 1-8, 1996.

[7] Vosahlik, J., Riding, K. A., Feys, D., Lindquist, W., Keller, L., Van Zetten, S., Schulz, B.: Concrete pumping and its effect on the air void system, Materials and Structures, 51(4), 94, 2018.

[8] Wang, L. K., Shammas, N. K., Selke, W. A., Aulenbach, D. B.: Gas dissolution, release, and bubble formation in flotation systems. Flotation technology. Humana Press, Totowa, NJ. 49-83, 2010.

[9] Rust, A. C., Manga, M.: Effects of bubble deformation on the viscosity of dilute suspensions. Journal of non-newtonian fluid mechanics, 104(1), 53-63, 2002.

[10] Krieger, I.M., Dougherty, T.J.: A mechanism for non -Newto suspensions of rigid spheres. Transactions of the Society of Rheology, 3(1), pp.137-152, 1959.

[11] Feys, D., De Schutter, G., Khayat, K. H., Verhoeven, R.: Changes in rheology of self-consolidating concrete induced by pumping. Materials and Structures, 49(11), 4657-4677, 2016.

[12] Feys, D., Roussel, N., Verhoeven, R., De Schutter, G.: Influence of air bubbles size and volume fraction on rheological properties of fresh self-compacting concrete. In 3rd International RILEM symposium on Rheology of Cement Suspensions such as Fresh Concrete. RILEM publications. 113-120, 2009.

The injection process – effecting mechanical properties?

B.Eng. Ludwig Hertwig, Prof. Dr.-Ing. Klaus Holschemacher
HTWK Leipzig, Structural Concrete Institute, Leipzig, Germany
Phone: 0341-3076 8832, e-mail: ludwig.hertwig@htwk-leipzig.de

Abstract

Injection as a novel manufacturing process for high-performance mortars in connection with fibres or textile reinforcement represents a powerful alternative to the pouring process. The subject of the investigations is the question if the use of the injection process compared to the pouring process not only provides better handling in the manufacturing process but also adds value in terms of the properties of hardened concrete. Subsequently, test specimens (cylinders with $d = 100$ mm, $h = 150$ mm and cubes with edge length $a = 150$ mm) were produced using pouring and injection process with the mixtures SCM 1, ECM 1a and LWSCM 1. The influence of the manufacturing process was assessed by the evaluation of hardened concrete properties like density, compressive strength, modulus of elasticity, water penetration depth and water absorption. The used formwork allowed the production of three test specimens on top of each other. This procedure enabled the investigation of various influencing factors as a function of the concreting height. The use of SLIPER (Sliding Pipe Rheometer) provided information on pumpability and reference values of rheological parameters to estimate injectability for the tested mixtures.

1 Introduction

The development and application of high-performance concretes and mortars[1] follows the desire for slim, lightweight and load-compliant design. For the production of such constructions, the following manufacturing processes are available to the user to date: Spraying, shotcreting, spinning, laminating and pouring [2].

Table 1 shows the manufacturing processes from the textile concrete sector where HPMs are commonly used. The application of the centrifugal process is

[1] High performance concrete is widely regarded as concrete with compressive strength of $f_{cm} \geq 60$ MPa [1]. Here, the concept „high performance mortar“ (HPM) combines the demands on high compressive strength, small maximum grain size $D_{max} \leq 2$ mm, high demand on workability and usability (architectural concrete).

limited to rotationally symmetrical and similar cross-sections in its range of applications. The spray, shotcreting and laminating process cannot guarantee an architectural concrete quality on all component surfaces, as not all sides can be produced smooth. Now, for the production of thin-walled, 3-dimensional building products in architectural concrete quality, the pouring process is of primary practical interest.

Table 1 Manufacturing processes in the textile concrete sector with possible component types, reinforcement and areas of application according to [2]

production techniques	component geometry	textile	component
laminate	2D components	2D textile, uncoated, degree of reinforcement almost arbitrarily controllable	facades, slabs
pouring	2D and simple 3D components	2D or 3D textile, uncoated and coated, reinforcement content limited by tightness of the structure, spacers necessary	facades, formwork elements, sandwich panels
shotcreting	2D and 3D components	2D textile, uncoated, degree of reinforcement almost arbitrarily controllable	plates in horizontal and vertical position, shells
centrifuging	circular hollow sections	2D or 3D textile, uncoated and coated, degree of reinforcement almost arbitrarily controllable	tubes, masts, piles

The production of complex and thin components by pouring is highly dependent on the rheological properties of the mixture. However, since self-compacting concretes and mortars are susceptible to segregation [3], the formation of the maximum flowability is risky. In addition, the process is time-consuming and labour-intensive, subjective application errors can easily occur.

The injection[2] process meets these challenges and therefore, it is an effective process for the production of high-performance mortars. The process uses the compression and distribution energy of the applied pump pressure. This may extend processing time significantly and reduces the cement and superplasticizer (SP) contents. Filling from below ensures that the fresh concrete level in the formwork rises evenly. The fresh concrete level increases constantly and thus

2 In the field of crack repair, the injection method describes the pressing of suspensions containing binding agents into concrete [4]. Since the pumping of concrete starts from a pipe diameter of $d \geq 80$ mm [4] in the field of construction engineering, the term injection is used to describe the analogy of filling complex geometries with HPMs under pressure.

the pump presses the matrix from below against flow barriers (reinforcement, geometry). The entire fresh concrete is in motion and thus spreads better.

The aim of the investigations is to find out whether the injection process improves, in addition to the advantages mentioned above in the fresh concrete sector, the properties of hardened concrete in comparison to the pouring process.

2 Case studies

2.1 Materials and test setup

For the injectability of mortars, there are currently no recommendations regarding the mix design. Therefore, it was assumed within the scope of these investigations that the mixing concept of a self-compacting mortar is target-oriented. Within the experiments, four different types of high-performance mortars are used (mix designs see Table 2):

- laboratory mixes: self-compacting mortar (SCM 1) and lightweight self-compacting mortar (LWSCM 1)
- industrial products: easy-compacting mortars (ECM 1, ECM 1a, ECM 2, ECM 2a) and grouts (G 1, G 2)

Table 2 Mix designs of used mortars showing masses, water-binder-ratio (*w/b*) and calculated density

mix	water	cement	fly ash	micro silica	agg.	SP	*w/b*	ρ_{fr}
			[kg/m³]				[–]	[kg/m³]
SCM 1	239	540	56	56	1198	4.0	0.38	2130
LWSCM 1	252	550	96	41	210	6.3	0.41	1160
ECM 1[b]	269	1923				14.1		2190
ECM 1a[a,b]	269	1923						2170
ECM 2	233	2000						2230
ECM 2a[a]	233	2000				5.0		2240
G 1[b]	288	1923						2210
G 2[b]	288	1923						2210

[a] additional dosage of SP [b] D_{max} = 1 mm
[c] no mix design available, sum of dry components

Table 3 Experimental program for comparison of pouring and injection regarding hardened concrete properties

measured properties	process	no.	mix	dimensions[a] component [mm]	quantity member [-]	total specimen [-]
ρ_d, f_{cm}, E_Y	pouring	1	*SCM 1*	480 x 100	3	9
		2	*ECM 1a*[b]	480 x 100	3	9
		3	*LWSCM 1*	480 x 100	3	9
	injection	4	*SCM 1*	480 x 100	3	9
		5	*ECM 1a*[b]	480 x 100	3	9
		6	*LWSCM 1*	480 x 100	3	9
$p, u_{wp}, u_{cap}, u_{cap,K}$	pouring	7	*SCM 1*	500 x 150 x 150	2	6
		8	*ECM 1a*[b]	500 x 150 x 150	2	6
		9	*LWSCM 1*	500 x 150 x 150	2	6
	injection	10	*SCM 1*	500 x 150 x 150	2	6
		11	*ECM 1a*[b]	500 x 150 x 150	2	6
		12	*LWSCM 1*	500 x 150 x 150	2	6

[a] cylinder (h x d) or cuboid (a x b x c) [b] workability for pouring/injection expected at lower limit

ρ_d density
f_{cm} compressive strength
E_Y Young´s modulus
p max. water penetration depth
u_{wp} water absorption (pressure)
u_{cap} water absorption
$u_{cap,K}$ water absorption according to *Karsten* [7]

To evaluate possible differences between the manufacturing process two normal concretes (SCM 1 and ECM 1a) and one lightweight concrete (LWSCM 1) are chosen. The workability of the easy-compacting mortar (ECM 1a – in conjunction with superplasticizer) is expected to be at the lower limit. Table 3 shows the experimental set up. Two different formworks produce two cuboid and three cylindrical components to be cut in each three cubes and cylinders. The set up allows showing differences between components, concreting heights, processes and mixes.

2.2 Test methods

For determination of possible changes of hardened concrete properties due to manufacturing process, concrete composition and height above lower level of fresh concrete level cylindrical and cuboid specimen were produced using mixes SCM 1, ECM 1a and LWSCM 1 (see Table 3). The treatment was the same procedure for all specimens[3]. The cylindrical specimens (see *Fig. 1*) delivered results about:

- Dry density (ρ_d) in compliance with DIN EN 12390-7.
- Compressive strength (f_{cm}) according to DIN EN 12390-3/B1. Loading speed increased to v = 2.4 kN/s. No grinding or adjustment of test specimens.
- Secant modulus of elasticity[4] (Young`s modulus – E_Y) according to DIN EN 12390-13. Usage of method B (stabilized modulus) with lower test stress $\sigma_p = 0.15 \cdot f_{cm}$. Accompanying specimen for initial test were prismen with a x b x c = 40 mm x 40 mm x 160 mm (average of 6 values). No grinding or adjustment of test specimens.

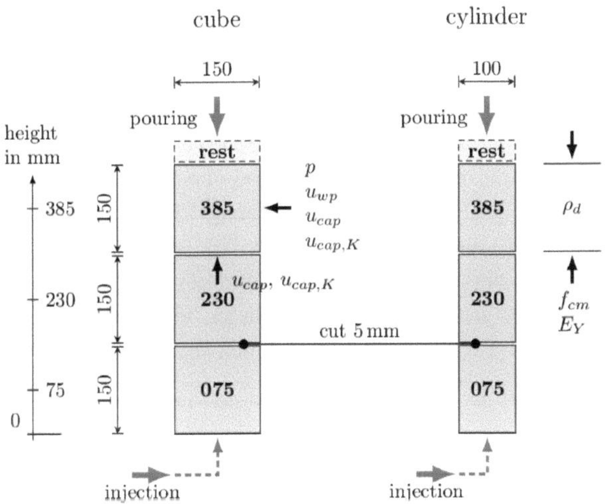

Fig. 1 Preparation and designation of specimen with their orientation during tests

The cuboid specimens (see *Fig. 1*) delivered results about:

[3] Components cut to cylinders with d x h = 100 mm x 150 mm or cubes with a = 150 mm and stored wrapped in foil for t = 28 d under climatic conditions (RH = 65 %, T = 20 °C).

[4] Due to the compact form (h/d = 1.5), the obstruction of the transverse strain and thus also the modulus of elasticity increases.

- Water resistance (maximum water penetration depth p in mm) in compliance with DIN EN 12390-08. Applying pressure $p = 5$ bar for $t = 72$ h. Water impinged side rotated 90° and the wooden formwork side.

- Water absorption during measurement of water resistance (u_{wp}) by weighing specimens before and after testing procedure of water resistance. Calculation based on dry mass in m.-%.

- Water absorption according DIN EN 772-11. Specimens prepared as oven-dried and tested under atmospheric conditions (u_{cap}) for $t = 24$ h. Therefore laying them 1 cm from bottom side up on spacers into water[5]. Testing and comparison[6] additionally due to water impinged side (wood, acrylic glass, cut side). Calculation of water absorption coefficient u_{cap} according to DIN EN 772-11, equation in section 8.1 in [6] and according to *Karsten* ($u_{cap,K}$) [7].

The prediction of needed rheological properties for successful injection is not possible. To evaluate the used mixes and provide recommendations the Sliding Pipe Rheometer (SLIPER) is used. It simulates the flowing process in an open pipe end [8]. The mortars are loaded in the pipe lifted to start position. Starting the measurement the pipe slides down and shears the lubrication layer around the concrete plug [8]. Repetitions at different load steps simulate various conveying rates. Time, distance and pressure data build a rheological model of the fluid. The parameters a and b of SLIPER describe the rheology of the fluid where a represents yield stress τ_0 and b plastic viscosity μ [8, 9]. Slump-flows[7] according to DIN EN 12350-8 put the measurements in context of one-point methods.

2.3 Production of the test specimens

For the production of the cuboid and the cylindrical specimens two types of formwork were necessary (see *Fig. 2*). The left shown formwork (for the cuboid components) consisted of impregnated wooden panels, an acrylic glass closed the front.

---

[5] Sides of specimen not impregnated.

[6] Subtraction of water absorption within first 5 min as suggested in [5].

[7] The slump-flow SF is carried out with a cone with bottom diameter $d = 200$ mm, top diameter $d = 100$ mm and a height $h = 300$ mm. Additionally the Haegermann cone is used (SF_{Haeg}).

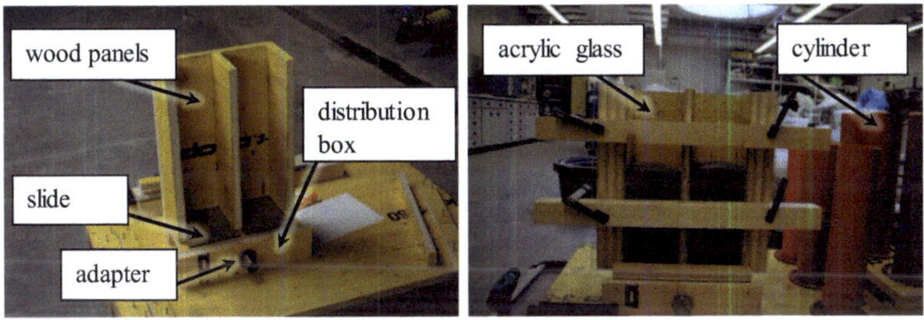

Fig. 2 Left: Empty formwork to inject from below and pour from top with slide and adapter of pump. Right: Filling of formwork during injection process (performing SCM 1)

During injection, the pump carried the concrete through the connected adapter into the distribution box. Inside this box the stream was diverted through holes (slides at position "open") with $d = 20$ mm equally upwards in both parts. In order to remove the adapter the slides had to be shifted to position "close". The pouring process used only the slide position "close". The formworks of the cylinders were plastic pipes, the only difference was the fact that the distribution box filled three pipes. The production of specimens followed this procedure:

- preparation of formwork
- intensive, multiple-stage mixing of the mix components with a compulsory mixer (capacity $V = 60$ l), mixing times $t = 10...15$ min
- injection: filling the mortars (SCM 1, ECM 1a, LWSCM 1, see Table 2) into pumping cone, injecting mortar steadily through adapter (pressure $p = 3...4$ bar at a flow rate of $Q \approx 0.2$ l/s) until reaching top of formwork
- pouring: filling the mortars (SCM 1, ECM 1a, LWSCM 1, see Table 2) steadily from top of formwork along panel side
- covering top with foil and leave curing for 2 days
- stripping the formwork, cutting and storing of the specimens (as described in *section 3.4*)

The production of the cuboid components by means of pouring and injection methods took place on different concreting days. Despite compliance with the mixing regime, deviations may occurred. Due to hot weather conditions, the temperature of the used mortars exceeded 25 °C.

3 Results

3.1 Comparison pouring - injection

The mean values of measured hardened concrete properties are shown in Table 4, additionally indicating manufacturing technique, height (location of the specimen during casting the concrete according to *Fig. 1*) and used concrete mix.

Table 4 Averages of all tests of hardened concrete using injection and pouring technique depending on mixture and specimen height h and mix

mix	h	ρ_d	f_{cm}	$E_{Y,m}$	p[c]	u_{wp}[c]	u_{cap}[c]	$u_{cap,K}$[c]
	[mm]	[kg/dm³]	[MPa]		[mm]	[m.–%]	[kg/m² h]	
pouring								
SCM 1	075	2.50	69.9	57 700		−0.042	0.31	0.37
	230	2.49	95.8	37 700		−0.090	0.92	0.16
	385	2.17	83.0	38 700		−0.082	0.94	0.72
ECM 1a[a,b]	075	2.86	67.5	35 300	8	−0.018		
	230	2.03	73.7	33 400	9	−0.086		
	385	2.04	72.6	34 700	14	−0.031		
LWSCM 1	075	1.99	30.7	11 400		−0.215	1.09	0.76
	230	1.11	41.9	12 800		−0.249	1.29	0.87
	385	1.99	33.3			−0.258	1.41	0.84
injection								
SCM 1	075	2.52	67.1	36 000	3	−0.031	0.66	0.57
	230	2.38	91.0	36 500	5	−0.137	0.94	0.72
	385	2.55	86.3	36 900	3	−0.145	0.56	0.51
ECM 1a[a,b]	075	2.05	50.6	40 600	7	−0.028		
	230	2.87	69.1	32 900	5	−0.125		
	385	2.98	55.2	33 300	7	−0.063		
LWSCM 1	075	1.86	29.4	11 700	3	−0.151	1.25	0.30
	230	1.91	36.5	12 000	2	−0.318	1.49	0.43
	385	1.81	34.5	11 700	2	−0.273	1.19	0.82

[a] additional dosage of SP [b] workability for pouring/injection expected at lower limit
[c] less than 3 specimens

The corresponding coefficients of variation (*CV*) in Table 5 give information about the scattering of test results.

The values of density (ρ_d) hardly showed any significant fluctuations in height, components, technique and within mixes. The highest coefficient of variation was $CV = 1.2$ % (ECM 1a, height = 230 mm, injection). This may result from the unsuitable mixture. Thus, cavities and pores formed on the outside during production. Overall, the high degree of agreement shows that the manufacturing processes do not give rise to differences in microstructure.

Table 5 Coefficients of variation of all tests of hardened concrete using injection and pouring technique depending on specimen height *h* and mix

mix	h	$?_d$	f_{cm}	$E_{Y,m}$	p^c	u_{wp}^c	u_{cap}^c	$u_{cap,K}^c$
	[mm]				[%]			
pouring								
SCM 1	075	0.0	6.6	15.4	-	0.4	1.7	1.7
	230	0.3	2.3	1.5	-	2.5	1.1	1.1
	385	0.1	12.1	8.7	-	7.9	1.1	1.1
ECM 1a[a,b]	075	1.2	4.2	1.2	37.5	-	-	-
	230	0.6	5.6	1.5	44.4	7.7	-	-
	385	0.6	3.6	3.5	3.4	95.6	-	-
LWSCM 1	075	0.3	8.8	-	-	2.7	-	-
	230	0.5	2.4	2.4	-	6.6	5.8	5.8
	385	0.6	12.1	-	-	0.2	-	-
injection								
SCM 1	075	0.3	14.9	2.7	66.6	0.1	-	-
	230	0.6	3.9	1.5	0.0	37.0	4.5	4.5
	385	0.6	5.7	1.0	0.0	31.6	8.5	8.5
ECM 1a[a,b]	075	1.1	9.1	33.6	28.5	75.1	-	-
	230	1.6	10.5	1.7	0.0	13.4	-	-
	385	0.5	13.0	3.8	-	18.1	-	-
LWSCM 1	075	0.3	21.6	6.6	0.0	38.1	3.1	3.1
	230	1.2	1.9	3.5	0.0	1.9	6.8	6.8
	385	0.5	4.5	1.8	50.0	5.5	11.4	11.4

[a] additional dosage of SP [b] workability for pouring/injection expected at lower limit
[c] less than 3 specimens

The strength properties of SCM 1 and ECM 1a are comparable, those of LWSCM 1 are lower. An influence of the manufacturing process on the mean compressive strength (f_{cm}) was apparent in the average of the values only for ECM 1a and LWSCM 1. The average value of SCM 1 for pouring and injection process was nearly the same (Δf_{cm} = 1 MPa). Evaluated according to the heights, the results for SCM 1 was different: In both methods, the lowest height had the lowest value, the middle height the highest and the highest height a middle strength (see boxplot *Fig. 3*). These tendencies also applied to the other mixtures. The largest scatters in the pouring process were at the upper height (CV = 12.1 %) and in the injection process at the lowest height (CV = 14.9 %). The mixes ECM 1a and LWSCM 1 showed higher strength (Δf_{cm} = 3 and 15 MPa) and lower fluctuations in the pouring process.

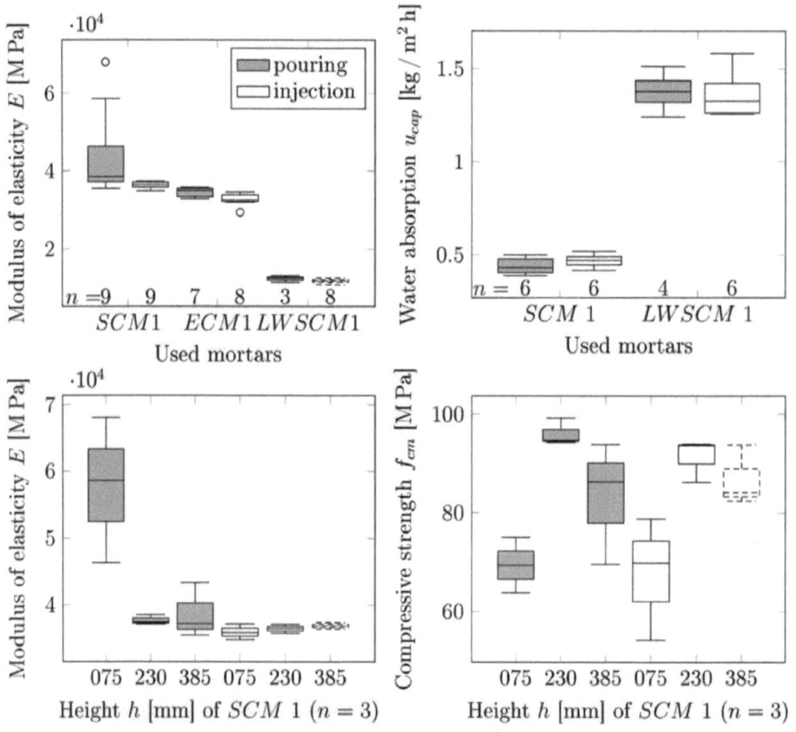

Fig. 3 Boxplots of average values by process, (partly) height and mix. Left: Modulus of elasticity. Right top: Water absorption. Bottom right: Compressive strength

In *Fig. 3* the influence of height and mix composition on modulus of elasticity, compressive strength and water absorption is shown. In this figure centre lines show the medians, box limits indicate the 25th and 75th percentiles, whiskers extend 1.5 times the interquartile range from the 25th and 75th percentiles, outliers are represented by dots and the sample size n is shown. With respect to the mixtures, the results of the modulus of elasticity tests were generally not so scattered (see boxplot *Fig. 3*). One exception was SCM 1 in the pouring process. The evaluation of SCM 1 according to the heights showed larger variations in the pouring process, especially for the lower and upper test specimens (CV = 15.4 and 8.4 %). The scatter in the injection method (average CV = 1.7 %) was lower, but also the E-moduli determined. LWSCM 1 performed in the same way, but fewer specimens were available[8]. The mixture ECM 1a showed better results in the pouring process. The reason for this was the high scattering of the lower test specimens in the injection process (CV = 33.6 %).

As the highest water penetration depth (p) was not visually detectable, for experiments no. 8 and 10 onwards (see Table 3) a fluorescent dye was used. Therefore, a direct comparison is only possible with mixture ECM 1a. The mean water penetration depth was lower with the injection method (Δp = 4 mm). The scattering was particularly high in the lower layers (CV = 66.6 % for SCM 1), with increasing height several times there was CV = 0 %[9].

Water absorption under pressure (u_{wp}) led to non-usable results. The change in specimen weight resulted from the superposition of the drying of the specimens (ambient conditions T = 28°C) and the increase in water under pressure during the tests. The effect of drying was predominantly. Therefore, the tests were not representative and the results are not discussed.

The results for water absorption under atmospheric conditions (u_{cap}) are shown in *Fig. 3*. A preliminary investigation indicated that the surface condition (wood, acrylic glass, cut side) after a time of t = 24 h has no significant influence on the water absorption. The values after 24 h according to DIN EN 772-11 were approx. 30 % higher than the calculation according to *Karsten* ($u_{cap,K}$). In general, the lightweight concrete (LWSCM 1) absorbed more water in the same time (see *Fig. 3*). For SCM 1, the water absorption in the injection procedure was slightly higher, but uniform. For LWSCM 1, the investigations showed the opposite. Due to the deviations in the upper test specimens (CV = 11.4 %), no clear evaluation of the manufacturing process is possible.

8 Some specimens showed cracks during the Young's modulus test and were therefore not included in the evaluation.

9 The accuracy of measuring the water penetration depth was ± 1 mm.

3.2 Rheology and injection

For each mix (SCM 1, ECM 1, LWSCM 1, ECM 2, G 1 and G 2, see Table 2) 4 load steps[10] with each 3 repetitions were applied. The investigations were carried out at fresh mortar temperatures $T = 25...28$ °C at times $t = 10...15$ min after initial water addition. Table 6 shows the results of the subsequent rheological investigations with the SLIPER and the single-point tests. The slump-flows according to Haegermann (SF_{Haeg}) stretched from $SF_{Haeg} = 130$ to 455 mm (ECM 2 and G 2). In general, the mixtures with a high spread had low yield stress ($a \approx \tau_0 = 0.25$ to 1.003 mbar). Only G 2 showed a high yield stress and plastic viscosity despite its good flow properties. The plastic viscosity (μ) is more widely distributed with good flow properties ($b \approx \mu = 0.811$ to 1.983 mbar h/m³). The measurement of a mixture later (SCM 1 after $t = 30$ min after measuring SCM 1 initially) led to a reduction of slump-flow ($\Delta SF_{Haeg} = 45$ mm) and an increase in the yield stress ($\Delta\tau_0 = 0.75$ mbar).

Table 6 Results of fresh concrete properties

mix	SF [mm]	SF_{Haeg} [mm]	$\Gamma_{P,SF(Haeg)}$ [-]	a [mbar]	b [mbar h/m³]
SCM 1	910	300	8.0	0.250	1.024
SCM 1$_{30min}$	800	255	5.4	1.003	1.006
LWSCM 1	520	260	5.8	0.542	0.811
ECM 1		140	0.9	1.784	1.769
ECM 1a[a]		260	5.7	0.876	1.983
ECM 2		130	0.6	2.672	1.233
ECM 2a[a]		185	2.4	1.844	0.861
G 1	815	260	5.8	0.685	1.480
G 2	850	455	19.7	2.486	1.985

[a] additional dosage of SP

SF slump flow; SF_{Haeg} slump flow (Haegermann cone)

$\Gamma_{P,SF(Haeg)}$ relative spread; a exp. value ≈ yield stress τ_0

b exp. value ≈ plastic viscosity μ

The comparison of the mixtures with and without additional superplasticizer[11] resulted in a superposition of: Increase of SF, decrease of yield stress, decrease (ECM 1a) and increase (ECM 2a) of plastic viscosity.

[10] The first step is without extra weight. Later extra weights of $m = 2.5$ kg, 5.0 kg and 7.5 kg were applied.

[11] The superplasticizer dosage chosen formed apparently robust mixtures.

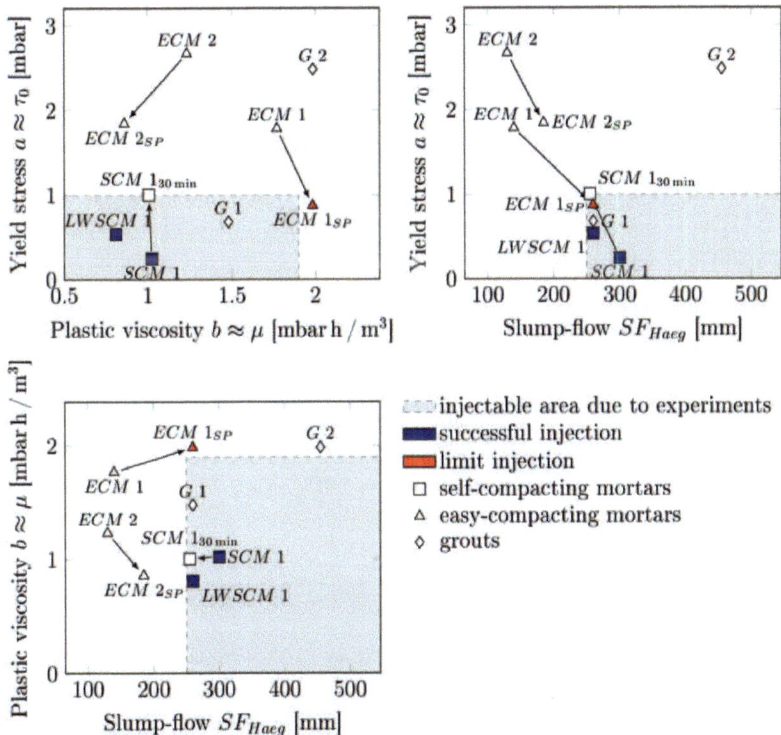

Fig. 4 Rheological properties of used mortars with estimation regarding ability of being injected regarding parameters *a* and *b* of SLIPER flow spread SF_{Haeg}.

The production of the test specimens showed that the mixtures SCM 1 and LWSCM 1 are easily injectable. ECM 1a marked visually the limit with $b = 1.983$ mbar h/m³ (with $\Delta b \approx 1$ mbar h/m³ to SCM 1). Considering the measured rheological properties, it is therefore possible to make statements on injectability. *Fig. 4* shows the measured parameters *a* and *b* as well as the spread of the mixtures. It also indicates the rheological changes during the addition of superplasticizers or the measurement at a later point in time. With ECM 1a as the limit, a zone[12] for injectable mixtures can be defined with respect to *a*, *b* and SF_{Haeg}. It is not possible to make a statement about injectability based on the flow spread. ECM 1a had a comparable flow spread, but a higher yield

[12] The limits regarding SF_{Haeg} and *a* are assumed in terms of high workability (see HPM).

stress and much higher plastic viscosity. The latter is mainly responsible for the injectability depending on the used pump.

4 Conclusion

The evaluation of the individual test variables does not show a clear tendency for or against one manufacturing process. However, this is also because some test methods, such as water penetration resistance, are not suitable for high-performance mortars and thus less significant values[13] were available.

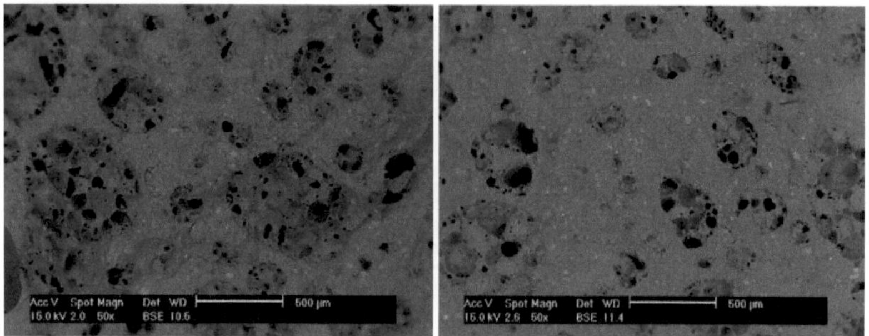

Fig. 5 Optical investigation of LWSCM 1 using REM-microscope. Left: Pouring. Right: Injection.

The pump pressure does not affect the structure of the matrices. This is demonstrated by the small deviations in the dry bulk density and the photographs of the lightweight concrete in *Fig. 5*. It indicates moreover no destruction of lightweight aggregates due to injection. The evaluations of f_{cm}, E_Y and u_{cap} generally show that injection method leads to unfavorable mean values of the hardened material properties, but also to lower scattering. In particular, the evaluation of the E-modulus shows that the injection process produces more uniform properties over the entire specimen height.

The Bland-Altman plot[14] (see *Fig. 6*) summarizes all measured values as pairs of the quotients of injection and pouring values together with their mean values. If, for example, the measured values of the injection method are higher, the value

[13] The measured penetration depths were too small for comparison. Imaginable ways to enlarge the penetration depths are higher water pressure or measuring earlier (e.g. after $t = 3$ d).

[14] The averages of E and f_{cm} are scaled down (factors 2500 and 10). Total count of data pairs $n = 124$.

pairs move upwards from the mean value of all value pairs. This is not the case. Most values are in the range of Mean Quot ± 1.96 · σ. This proves that the injection method and the pouring method are equivalent for suitable compounds in terms of properties. Taking into account the lower scattering, the injection process shows its performance capability.

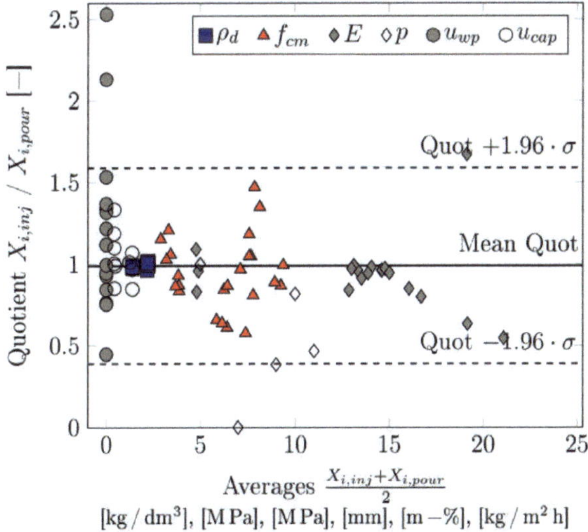

Fig. 6 Bland-Altman plot of all measured concrete properties using injection or pouring technique.

The tests point out that the rheological parameters of yield stress and viscosity highly influence injectability. It appears that plastic viscosity has a greater influence than yield stress. The pump pressure overcomes the yield stress more easily.

The injection process is an efficient alternative, especially in comparison to the pouring process. Reproducibility increases and in this fashion, the quality of thin-walled concrete components is ensured. The applied pump pressure improves the workability of the matrix and distributes it evenly. In addition, the reduction of labour costs is a significant economic advantage.

The following further investigations could aim at:

- quantification of necessary fresh concrete properties for an estimation of injectability regarding geometry and reinforcement
- influence of the tribological properties of the formwork the procedure
- influence of the heat development of the screw conveyor on the mixture

Literatur

[1] König, G.; Tue, N. V.; Zink, M.: Hochleistungsbeton. Bemessung, Herstellung und Anwendung. Ernst, Berlin, 2001.

[2] Hegger, J.; RWTH Aachen: Kompetenzzentrum Textilbeton. http://www.textilbeton-aachen.de/information/herstellverfahren-fuer-textilbeton/ (Abruf: 16.08.2018).

[3] Lowke, D.: Sedimentationsverhalten und Robustheit Selbstverdichtender Betone. Dissertation, TU München, München, 2013.

[4] Peter, N. K.: Lexikon Bautechnik. 15.000 Begriffsbestimmungen, Erläuterungen und Abkürzungen. Müller, Heidelberg, 2005.

[5] Haindl, K.; Schöner, T.; Zirkelbach, D. et al.: Was ist bei Karsten & Co. zu beachten? Bauen im Bestand : B + B 39 (2016), Heft 3, S. 33–37.

[6] Deutsches Institut für Normung: DIN EN 772-11:2011-07- Prüfverfahren für Mauersteine - Teil 7: Bestimmung der kapillaren Wasseraufnahme von Mauersteinen aus Beton, Porenbetonsteinen, Betonwerksteinen und Natursteinen sowie der anfänglichen Wasseraufnahme von Mauerziegeln. Beuth Verlag, Berlin, 2011.

[7] Karsten, R.: Bauchemie. Ursachen, Verhütung und Sanierung von Bauschäden. Handbuch für Studium und Praxis. C. F. Müller, Heidelberg, 2003.

[8] Kasten, K.: Gleitrohr-Rheometer. Dissertation, TU Dresden, Shaker-Verlag, Aachen, 2009.

[9] Secrieru, E.; Fataei, S.; Schröfl, C. et al.: Study on concrete pumpability combining different laboratory tools and linkage to rheology. Construction and Building Materials 144 (2017), S. 451–461.

Adhesive mortars properties: Squeeze Flow and Contact Visualization

Alessandra Lie Fujii-Yamagata[1], Fábio Alonso Cardoso[2], Anne Daubresse[3], Evelyne Prat[3], Mohend Chaouche[1]

[1] Laboratoire de Mécanique et Technologie, École Normale Supérieure Paris-Saclay/ CNRS/ UMR8534, Cachan, France

[2] Department of Construction Engineering, Escola Politécnica, University of São Paulo, São Paulo, Brazil

[3] Centre d'Innovation Parexgroup, St Quentin Fallavier, France

Phone: +1 (720) 593 2312, e-mail: alesandra.yamagata@gmail.com

Abstract

The present study evaluates adhesive mortars squeeze flow behavior and contact generation for formulation with different cellulose ether (CE) content. To predict the behavior of the material during tile application, squeeze flow technique was performed after 0, 10 and 20 minutes of resting. Then, to evaluate the consequent contact generated, small depth-of-field optical microscopy was used to visualize the areas of contact for different waiting times (5 and 20 minutes). The results showed that squeeze flow behavior can influence the contact generation between mortar and tiles. Also, contact visualization was able to verify different zones of contact due to skin break.

1 Introduction

Adhesive mortar is a masonry element used to attach tiles. During application, the mortar is applied to the substrate with a trowel in a wide surface and then the tiles are put onto it. It is a simple procedure, but it requires complex rheological behavior during tile placement and the generation of contact. When the tile is being placed, it will squeeze the adhesive mortar, demanding from a good squeeze flow behavior of the material, which means proper spread as it is compressed. Then, it should generate large contact extension to provide favorable conditions for adhesion to the tile.

Squeeze flow is a measurement technique that consists on compressing a cylindrical sample within two parallel plates. The technique is widely used on other fields as cosmetics, ceramics, food and composites. The large range of application is related to the absence of many problems found in rheological measurements, as loss of contact between the material and the shearing element, fibers entanglement, capillary clogging and restrictions due to torque limitations [1]. Its use is becoming more frequent as an alternative/complementary method for cement-based pastes and mortars, especially to simulate flow situations associated to geometric restrictions (extrusion, spreading, brick laying, flow through a nozzle during pumping or spraying) [2–5]. This technique can offer valuable information regarding the mortar's behavior during tile placement.

After the tile is placed and properly squeezed, the contact that is made between the tile and the mortar becomes one of the greatest issues to guarantee good adhesion of the tile though. In this system, the interface between the tile and the mortar is most fragile part, where most of problematics occurs, so tackling the contact generation is key to understand adhesion. Recent studies have been tackling problematics such as skin formation, which may interfere in the contact generated between mortar and tile. However, the evaluation of the contact generation is still very limited, mainly evaluated through glass visual contact generated or final adhesive stress. Zurbriggen et al suggested a modification to European standards EN1346 and EN1347 to observe the real areas of contact by applying an incident light [6].

In this context, the present study employs Squeeze flow technique to evaluate mortar's squeezing behavior and Optical Microscopy to assess contact generation. Formulations with various CE content were used to obtain mortars with different behaviors.

2 Materials and Methods

2.1 Materials and Formulations

This work employed basic compositions of adhesive mortars with CEM I 52,5 N CE CP2 NF from Lafarge, silica sand PE2LS from FULCHIRON, Redispersible Polymer Powder from Momentive and organic viscosity modifier (CE: hydroxyethyl methyl cellulose with 40000 mPa.s) from Dow Chemicals. The formulations used in this study are shown in Table 1. Each adhesive mortar formulation was mixed in a planetary mix according to EN-196-1.

2.2 Squeeze Flow

Squeeze flow measurement was done in a rheometer MCR 301 from Anton Paar with a cross hatched geometry with a diameter of 50 mm and an aluminium cup of 52 mm.

The mortar samples were applied with the help of an aluminum tool in a cylindrical shape of 6 mm height and 40 mm diameter. Adhesive mortars consistency tends to be "stick" and hard to mold, thus, usual rendering mortars molding techniques are not efficient and causes deformed samples. Some adaptation to squeeze flow have been used by some authors. Costa et al. [7] used a cylindrical tube as mold that are many pieces assembled, during the test the pieces are able to slide with the mortar. Kudo et al. [8] used a cylindrical plastic mold that is fixed during the entire test, the puncture touches the sample and pull out the material with the wall, similar to a surface tension measurement of a liquid. The molding technique developed in this investigation prepares a uniform shape of the samples, allowing free squeeze without any constraints. The technique is based on a metallic comb that turns to form the cylindrical sample as illustrated in Figure 1.

Table 1: Formulations of adhesive mortars used (wt%)

Formulations	CEM I	Sand	Latex	CE	Water/powder	Entrained air
0.1% CE	30%	67.40%	2.5%	0.1%		13%
0.25% CE	30%	67.25%	2.5%	0.25%	0.34	20%
0.4% CE	30%	67.10%	2.5%	0.4%		22%

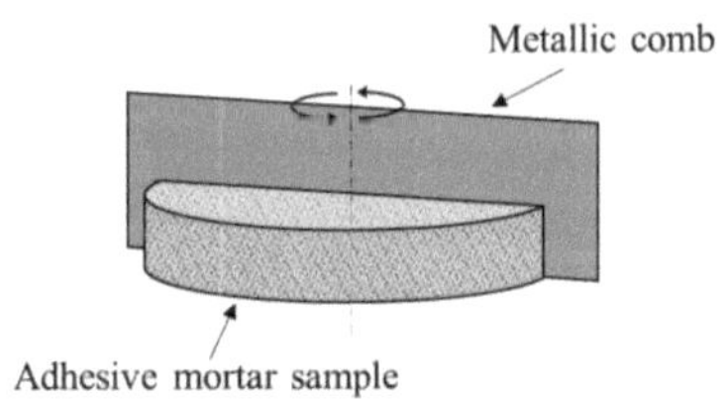

Figure. 1: Illustration of adhesive mortar sample preparation: the metallic comb turns around axis to form the cylindric shape of the sample

The samples were put to rest for different waiting time: 0, 10, and 20 minutes. After this time, the sample was put in the rheometer for the squeeze test, where it was compressed at an imposed velocity of 1 mm/s for 3 mm.

In the test, the sample is compressed at constant velocity and the normal force is measured. In Figure 2, a typical squeeze flow curve of a mortar shows three stages of behavior: Stage I: small strain — elastic deformation; Stage II: moderate strain — plastic deformation or viscous flow, where the material can deform considerably with little increase in the applied force; and Stage III: generally associated to large strains — the force required to squeeze the material increases substantially, which is known as the strain hardening stage [9]. There are conditions where Stage II may not occur, for example, in cases where phase segregation occurs from the beginning of the test and induce great load increase.

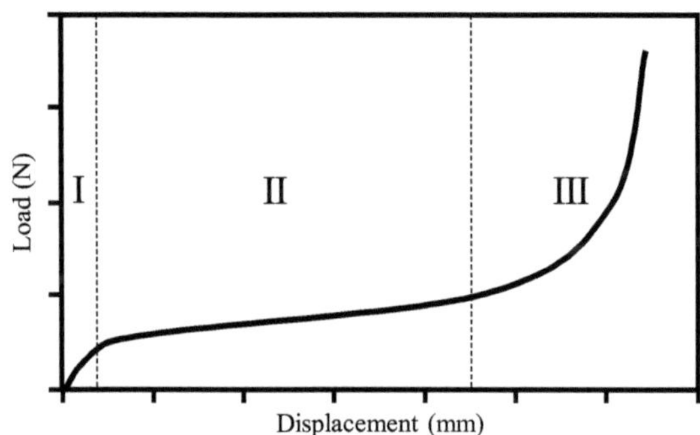

Figure. 2: Typical load vs. displacement curve of a displacement-controlled squeeze flow test, illustrating the three main stages of the material's behavior [9].

2.3 Small depth-of-field Optical Microscopy

To visualize the contact between tile and mortar with high definition, we developed a microscopical technique. It is based on a simple concept of lens focusing. Using a microscopical set-up with a small depth of field, it is possible to visualize only a specific focus distance where the interface is located.

In this work, a Digital Microscope Keyence VHX with VH-Z100R lens was used; the set-up is shown in Figure 3. The adhesive mortars were applied with a toothed comb on 5 cm x 5 cm x 5 mm prexiglass tiles and different waiting times (5 and 20 minutes), another tile was put with a 2 kg weight on the top for 30 seconds. The 2 Kg weight after different waiting time was used as defined by EN 1346 – Determination of open time [10].

After application, the samples were divided in different zones for the sample analysis, also divided by the areas where the contact was generated by the compression from the top of the ribs (rib) or the material that squeezed and flowed into between the ribs (between ribs).

The environmental conditions were kept at 23°±1C and 40-50% RU. To improve reliability, each study sample preparation was done on the same day and one after another. This ensured that the environmental conditions of the room between the formulations of the same study were similar, but discrepancies could occur between the two studies samples.

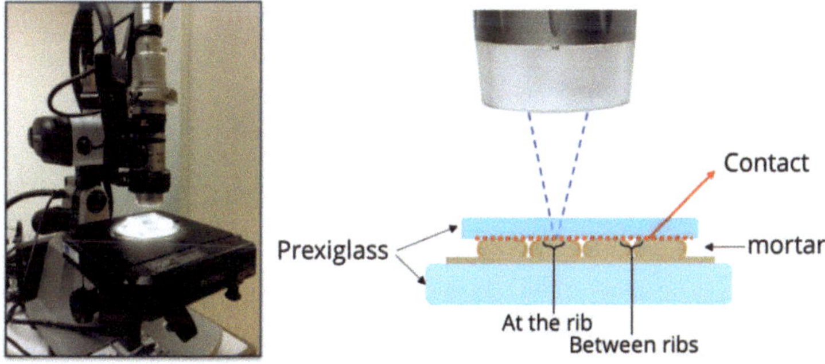

Figure. 3: Digital microscope VHX (Left) and scheme of image obtained (Right).

4 Results

4.1 Squeeze Flow

In Figure 4, squeeze flow test results are shown for the formulations with different CE contents at waiting times.

For a waiting time of 0 minutes, as the CE content is increased, the normal force required to compress the sample also increased. This is related to the CE thickening behavior. For a waiting time of 10 minutes, the normal forces evolve, however, it clearly shows how the formulation with 0.1% CE content has a faster evolution, overpassing the formulation with 0.25% CE. Finally, with a longer waiting time of 30 minutes, the formulation of 0.1% normal forces increased, becoming even higher than the formulations with 0.4% CE.

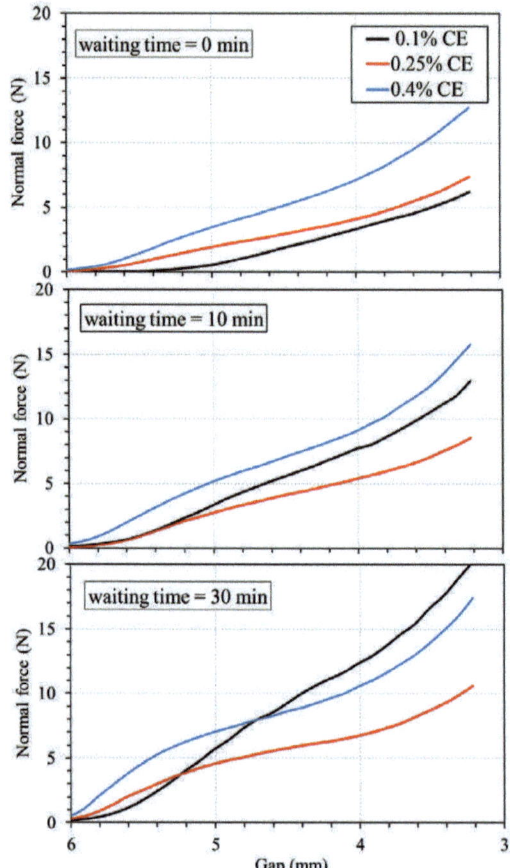

Figure. 4: Squeeze flow curves of formulations with different CE content at different waiting time (0, 10 and 20 minutes).

The causes could be related to faster structure buildup of the materials, entrained air, as well as possible phase separation during squeeze [2]. For the formulation with 0.1% CE, a faster structuring kinetics may induce a faster increase of squeeze forces as the material rests. Air entrainment could play a role, since stabilized bubbles of the formulations with higher CE content can generate lower squeezing forces independently of the material structuring. Additionally, two types of phase separation seem to be involved in the squeeze results. Firstly, bleeding probably influenced the formulation 0.1% CE at 0 minutes that showed very small forces at the beginning of the test. Then, for longer waiting times of

10 and 20 minutes of waiting, the curves start at a smaller value than the other formulations, followed by an inversion of the results, which indicates strain hardening due to solid/liquid separation induced during squeezing. CE has a thickening effect by increasing water viscosity, which is effective in reducing phase separation effect. The content of 0.1% of CE does not seem to be enough to avoid phase separation and, consequently, the effects of the heterogeneous flow negatively influenced the squeezing performance of the mortar.

High squeezing forces may indicate a material that cannot easily deform to generate good contact with the tile. The technique has a high potential for being applied to quality control as well as to studies regarding the evolution of the rheological behavior of mortars.

4.2 Contact Generation

The small depth-of-field optical microscopy technique was used to observe the contact between mortar and the tile during fresh state. The images were taken of formulations with different CE content after 5 and 20 minutes of waiting times, at two different zones: at the rib and between ribs. From the images, it was possible to observe different types of contact were observed.

In Figure 5a, the formulation 0.1% CE with lower CE content is shown. The images clearly show a great loss of contact from 5 minutes waiting time to 20 minutes waiting time. Indicating that as the mortar evolves during resting, the contact is greatly affected. This contact was also shown to be fairly homogenous throughout the sample, and at the different zones, no different was observed. After 20 minutes, due to very poor squeeze, no zone between ribs were observed.

In Figure 5b and 5c, the formulations with 0.25% CE and 0.4% CE are shown. Very different contact is observed at the different zones of the sample due to skin formation. Skin formation was measured by Fujii et al [11], showing a more rigid skin at the surface of the formulations with higher CE content. In the region at the ribs, the formulation with higher CE seem to have a more intense interfacial evolution, resulting in faster loss of contact. However, at the region between the ribs, it seems that the material with higher CE content is able to generate better contact.

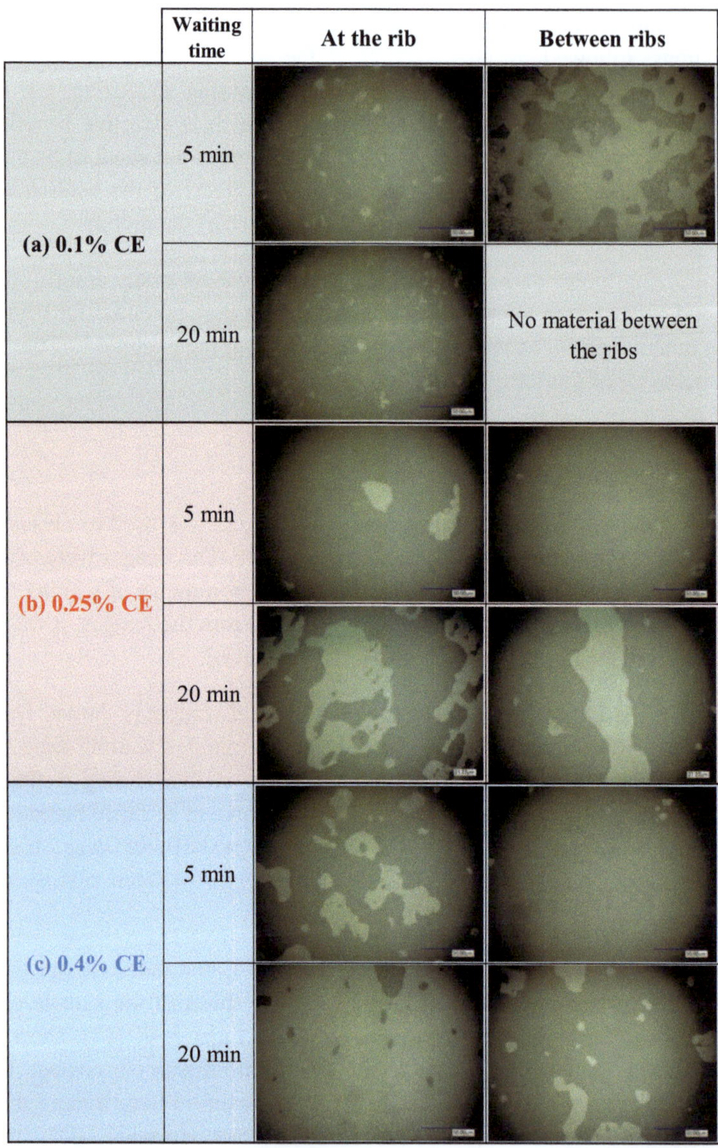

Figure. 5: Contact between mortar and tile after 5 and 20 minutes of resting at two different zones (at the ribs and between ribs) for the formulations (a) 0.1% CE, (b) 0.25% CE, and (c) 0.4% CE.

Overall, it is shown that the material with as low as 0.1% CE content is not able to generate good contact. This result can be related to the squeeze flow behavior of the samples. All samples were prepared at a constant force of 2kg for 30 seconds after the tile is placed, therefore, squeeze flow behavior play a considerable role on the contact generation. From the images and squeeze flow results, the formulation with lower CE content showed a good contact and squeeze flow behavior in the beginning for small waiting times, however, as the waiting time increased, this behavior changed, and the material no longer was able to be squeezed properly, release fresh material, and produce a good contact.

For higher CE content, the evolution of the squeeze seems lower, but there was a skin formation at the surface of the material. Contact results, however, despite this skin, as the material maintained better squeeze behavior, it was able to be compressed and form a better contact between the mortar and tile.

5 Conclusions

The present study proposed a rheological evaluation of adhesive mortars with very simple techniques, but very powerful to measure material's behavior, which can affect the performance of the tiling system. Squeeze flow technique have shown to be powerful to evaluate how the rheological properties of the material evolve as the waiting time is increased. For this research, a critic change of behavior for the formulation with lower CE content was observed.

Then, small depth-of-field optical microscopy was used to evaluate the contact generation between the mortar and tiles. The technique was able to identify different types of contact and also follow their evolution as the waiting time is increased. Through the results, one can notice that despite the skin formation, good contact was observed with formulations with higher CE content. At the same time, squeeze flow behavior had a role on reducing the contact generation of the formulation with 0.1% CE even though there was no skin formed.

6 Acknowledgements

The authors acknowledge ParexGroup for financing this project. CNPq – Brazil for F.A. Cardoso's post-doc grant.

Bibliography

[1] Cardoso, Fábio A.: Method of formulation of rendering mortars based on granulometric distribution and rheological behavior, PhD Thesis, University of São Paulo, 2009 (in Portuguese)

[2] Cardoso, Fábio A.; Fujii, Alessandra L.; Pileggi, Rafael G.; Chaouche, Mohend: Parallel-plate rotational rheometry of cement paste: Influence of the squeeze velocity during gap positioning, Cem. Concr. Res. 75, 66–74, 2015

[3] Toutou, Zahia; Roussel, N.; Lanos, C.: The squeezing test: A tool to identify firm cement-based material's rheological behaviour and evaluate their extrusion ability, Cem. Concr. Res. 35, 1891–1899, 2005

[4] Phan, T.H.; Chaouche, Mohend: Rheology and stability of self-compacting concrete cement pastes, Applied Rheology, 15, 2005

[5] Collomb, J.; Chaari, F.; Chaouche, M.: Squeeze flow of concentrated suspensions of spheres in Newtonian and shear-thinning fluids, J. Rheol., 48, 405–416, 2004

[6] Zurbriggen, Roger; Herwegh, M.; Pieles, U.; Huwiler, L.: A new laboratory method to investigate skin formation and Open Time performance, in: Third Int. Drymix Mortar Conf. Idmmc Three, Nürnberg, Germany, 2011

[7] Costa, M.R.M.M.; Pereira, E.; Pileggi, Rafael G.; Cincotto, Maria A.: Study of the influential factors on the rheological behavior of adhesive mortar available in the market, Ibracon Struct. Mater. J. 6, 399–405, 2013

[8] Kudo, E.; Cardoso, Fábio A.; Pileggi, Rafael G.; Squeeze flow applied to adhesive mortars: influece of experimental parameters of configuration and displacement rate, An. Do IV SBTA, Belo Horizonte, 2011 (in Portuguese)

[9] Cardoso, Fábio A.; John, Vanderley M.; Pileggi, Rafael G.; Rheological behavior of mortars under different squeezing rates, Cem. Concr. Res. 39, 748–753, 2009

[10] CEN, EN 1346: Adhesives for tiles — determination of open time, Eur. Stand, 2007

[11] Fujii, Alessandra L.; Cardoso, Fábio A.; Chaouche, Mohend: Interfacial rheology : measurement technique for adhesive mortar-air interface, Greim, Markus, 26th Conf. Work. Rheol. Build. Mater., Regensburg, 2017

Rheologische Untersuchungen an Frischbetonmischungen für Spritzbeton

Maria Thumann[1)], Lukas Briendl[2)], Joachim Juhart[2)], Wolfgang Kusterle[1)]
[1)] Fakultät Bauingenieurwesen, Labor für Betontechnologie, OTH Regensburg
[2)] Institut für Materialprüfung und Baustofftechnologie, TU Graz
Tel.: 0941-943 1200; e-mail: maria1.thumann@oth-regensburg.de

Kurzfassung

Im Forschungsprojekt „ASSpC – Advanced and Sustainable Sprayed Concrete" werden neue Spritzbetonrezepturen mit einer erhöhten Dauerhaftigkeit und geringen Umweltauswirkung entwickelt. Bei der Rezepturentwicklung ist insbesondere zu berücksichtigen, dass die Mischungen auch die Anforderungen an eine ausreichende Verarbeitbarkeit und Frühfestigkeitsentwicklung erfüllen. Im Forschungsprojekt wird untersucht welche Parameter die Pumpbarkeit der Frischbetonmischungen für Nassspritzbeton beeinflussen und wie die Pumpbarkeit messtechnisch beurteilt werden kann. Hierzu werden zahlreiche Prüfmethoden zur Beurteilung der Konsistenz, der rheologischen Eigenschaften und der Stabilität der Mischungen ausgewählt und systematisch bei Labor- und Großspritzversuche eingesetzt. In diesem Beitrag werden erste Ergebnisse der Versuchsreihen vorgestellt.

1 Einleitung

Die Verarbeitbarkeit von Nassspritzbeton ist ein wesentlicher Bestandteil der Rezepturentwicklung und Voraussetzung für den erfolgreichen Einsatz von Spritzbeton in der Praxis. Dabei ist insbesondere auf eine gute Pumpbarkeit und Spritzbarkeit der Mischungen zu achten. Im Rahmen des Forschungsprojektes „ASSpC – Advanced and Sustainable Sprayed Concrete" wird die Entwicklung von Spritzbetonrezepturen hinsichtlich einer erhöhten Dauerhaftigkeit und geringen Umweltauswirkung fokussiert. Dabei werden auch die Anforderungen an die Verarbeitbarkeit und Frühfestigkeitsentwicklung untersucht. Schwerpunkt in diesem Beitrag liegt auf den rheologischen Untersuchungen von Frischbetonmischungen für Nassspritzbeton.

2 Verarbeitbarkeit von Frischbetonmischungen

Die Verarbeitbarkeit von Frischbeton umfasst zahlreiche Eigenschaften [1]. Bei Spritzbeton ist insbesondere die Pumpbarkeit der Frischbetonmischung ein wesentlicher Bestandteil für die Beurteilung der Verarbeitbarkeit. Der Beton muss ohne Änderungen in der Mischungszusammensetzung gefördert werden und unter Druck ausreichend stabil und fließfähig sein. Bei der Förderung von Beton in einer Rohrleitung bildet sich an der Rohrwandung eine sogenannte Gleitschicht aus (siehe Bild 1). Sie besteht aus Zement, Wasser, ggf. Zusatzstoffen, ggf. Zusatzmitteln und den Feinteilen der Gesteinskörnung. Der Dicke und der Zusammensetzung der Gleitschicht kommt bei der Förderung in Rohrleitungen eine entscheidende Bedeutung zu. Sie reduziert den Förderwiderstand in der Leitung und der Beton wird im restlichen inneren Querschnitt als Pfropfen gefördert [2; 3].

3 Prüfverfahren zur Beurteilung der Verarbeitbarkeit von Frischbetonmischungen

3.1 Einpunktprüfverfahren

Die Beurteilung der Pumpbarkeit von Frischbetonmischungen erfolgt bei Nassspritzbeton in der Regel anhand von Kennwerten zur Beurteilung der Konsistenz, z. B. dem Ausbreitmaß oder dem Setzmaß [4]. Die Verfahren sind für die Überprüfung der Konsistenz der Mischungen auf der Baustelle geeignet, jedoch nicht für die zuverlässige Beurteilung der Pumpbarkeit von Frischbetonmischungen. In zahlreiche Veröffentlichungen wird beschrieben, dass es keinen eindeutigen Zusammenhang zwischen der Konsistenz und der Pumpbarkeit von Frischbeton gibt [2; 5–8].

Bei den nachfolgenden Versuchsreihen wurde das Ausbreitmaß [9] bei allen Frischbetonmischungen geprüft, um diese Erkenntnisse zu bestätigen. Des Weiteren sollen Rückschlüsse im Vergleich zu rheologischen Prüfverfahren und zu Prüfverfahren zur Beurteilung der Stabilität ermöglicht werden.

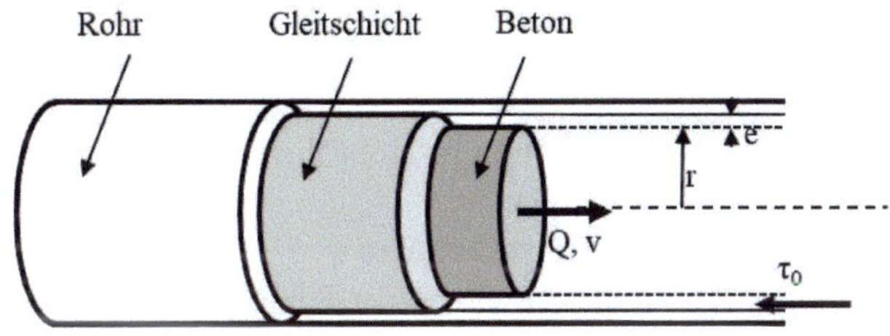

Bild 1: Förderung in einer Rohrleitung nach [2]

3.2 Rheologische Prüfverfahren

Für die Beurteilung der rheologischen Eigenschaften der Frischbetonmischungen hinsichtlich der Pumpbarkeit wurde das Gleitrohr-Rheometer „Sliper" [10] gewählt (siehe Bild 2). Das „Sliper" wurde sowohl für den Einsatz im Labor als auch auf der Baustelle entwickelt.

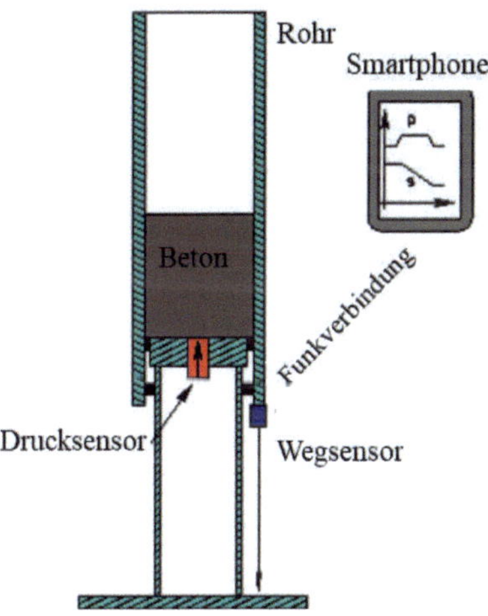

Bild 2: Systemskizze Gleitrohr-Rheometer Sliper [10]

Das Gleitrohr-Rheometer besteht aus einem aufgeständerten Kolben und zwei verbundenen Rohrteilen, die entlang des Kolbens geführt werden. Für den Messvorgang wird das Rohr mit Frischbeton gefüllt und durch Lösen einer Arretierung bewegt sich das Rohr nach unten. Dabei werden sowohl der Druck als auch die Geschwindigkeit aufgezeichnet. Durch Zusatzgewichte, die am Gleitrohr-Rheometer angebracht werden, können unterschiedliche Geschwindigkeiten und Drücke erreicht werden. Mithilfe eines hinterlegten Berechnungsmodells werden die Daten in Pumpprognosen umgerechnet und in einem Förderdruck zu Fördermengen Diagramm (p-Q Diagramm) dargestellt. Daraus können die rheologische Kennwerte zur Beurteilung der Pumpbarkeit ermittelt werden. Der Beiwert a steht dabei für die Fließgrenze und der Beiwert b für die Viskosität, ähnlich wie bei einer Fließkurve. Die Beiwerte wurden gezielt gewählt, um eine Verwechslung mit den Bezeichnungen aus der Rheologie zu vermeiden [2]. Die mit dem Sliper ermittelten Kennwerte von Viskosität und Fließgrenze können unter Berücksichtigung folgender Parameter in einen Betondruck umgerechnet werden: Betondichte, Rohrdurchmesser, Förderhöhe, Fördermenge und Förderlänge. Die Ergebnisse der Slipermessungen zur Beurteilung der Pumpbarkeit sind nicht vergleichbar mit den Ergebnissen anderer Rheometer, welche die rheologischen Eigenschaften der gesamten Betonmischung prüfen.

3.2 Prüfverfahren zur Beurteilung der Stabilität

Für die Pumpbarkeit ist eine ausreichende Stabilität der Frischbetonmischungen unter Druck erforderlich, damit eine Entmischung in der Förderleitung nicht zu Blockaden führt. Für die Beurteilung der Stabilität der Mischungen wurde das Prüfverfahren nach dem öbv-Merkblatt „Weiche Betone“ [11] mit der Betonfilterpresse gewählt. Das Verfahren wurde entwickelt, um die Stabilität von Betonen mit einem Ausbreitmaß ≥ 560 mm beurteilen zu können.

Ein Behälter mit einem Volumen von 10 l wird mit einem Druck von 3 bar beaufschlagt und über eine Austrittsöffnung am Boden des Behälters wird die Menge an Filtratwasser aufgefangen (siehe Bild 3).

Nach 15 min und 60 min wird die Filtratwassermenge bestimmt und das Ergebnis wird in kg/m³ zu beiden Prüfzeitpunkten angegeben.

$$\text{Filtratwasser [kg/m}^3\text{]} = \frac{\text{Wasserabgabe [ml]} \cdot 1000}{\text{Gefäßvolumen [cm}^3\text{]}}$$

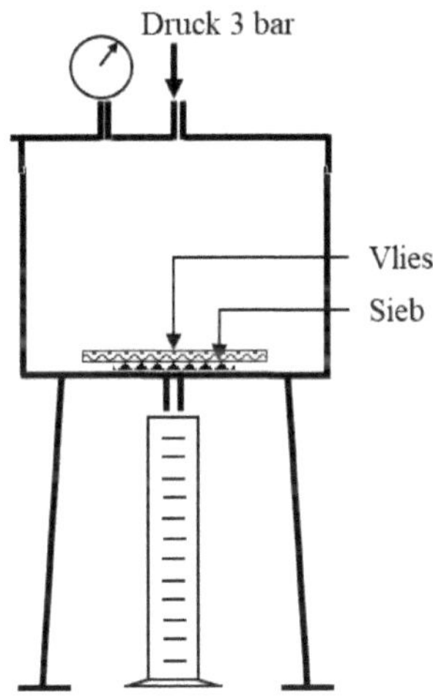

Bild 3: Schemaskizze Filterpresse nach [11]

4 Versuchsprogramm Labor

4.1 Referenzrezeptur

Für die Laborversuche wurde die in Tabelle 1 aufgeführte Referenzrezeptur gewählt. Auf dieser Grundlage wurden die Mischungsvariationen für weitere Versuchsserien entwickelt, z. B. Variationen im Leimgehalt oder Wassergehalt.

Tabelle 1: Referenzrezeptur Laborversuche

	Dichte	Masse	Volumen
	[kg/dm³]	[kg/m³]	[dm³/m³]
CEM I 52,5 N SR0	3,2	428	131
Wasser	1,0	214	214
Gesteinskörnung Quarz 0/8	2,7	1655	625
Luftgehalt	---	---	30
Gesamt	---	2397	1000

Es wurde quarzitische Gesteinskörnung mit einem Größtkorn von 8 mm verwendet. Die Sieblinie entspricht den Anforderungen der öbv-Richtlinie Spritzbeton [12] und wurde aus 7 Kornfraktionen zusammengestellt, um jegliche Streuung zu vermeiden.

4.2 Versuchsergebnisse

4.2.1 Leimvolumen

Das Leimvolumen beeinflusst die Ausbildung der Gleitschicht an der Rohrwandung und damit die Pumpbarkeit. Die Pumpbarkeit von Mischungen mit unterschiedlichen Leimgehalten wurde in Laborversuchen mithilfe des Slipers untersucht. Die Zusammensetzung der Gleitschicht, insbesondere die maximale Korngröße, beruht derzeit auf Annahmen. In der Literatur finden sich dazu keine einheitlichen Aussagen. Zum theoretischen Leimvolumen gehören bei den nachfolgenden Serien der Zement, die Zusatzstoffe, das Wasser, die Feinteile der Gesteinskörnung < 0,125 mm und der Luftgehalt.

Ausgehend von der Referenzmischung wurde das Leimvolumen bei der Mischung LV_n (**L**eim**v**olumen_**n**iedrig) um ca. 30 l verringert (siehe Tabelle 2).

Tabelle 2: Volumenanteile von Rezepturen mit unterschiedlichem Leimvolumen

	Einheit	Referenz	LV_n	LV_n_LP
CEM I 52,5 N SR0	[dm^3/m^3]	131	118	118
Wasser	[dm^3/m^3]	214	192	192
Gesteinskörnung < 0,125 mm	[dm^3/m^3]	37	40	40
Luftgehalt	[dm^3/m^3]	36	38	76
Gesamt Leimvolumen	[dm^3/m^3]	419	388	426

Versuchsergebnisse der Slipermessung zeigen, dass eine Verringerung des Leimvolumens zu einer Erhöhung des Förderdrucks führt (siehe Bild 4).

Das Leimvolumen kann jedoch durch die Einführung künstlicher Luftporen erhöht werden. Mit der der Zugabe eines Luftporenmittels konnte der Leimgehalt der Mischung LV_n_LP (**L**eim**v**olumen_**n**iedrig_**L**uft**p**orenmittel) im Vergleich zur Rezeptur LV_n um 38 l erhöht werden. Die Erhöhung des Leimvolumens führte zu einer Reduzierung des Pumpendrucks und die Pumpprognose entspricht in etwa der Referenzmischung. Der Unterschied des Leimvolumens der Referenzmischung zur Rezeptur LV_n_LP beträgt lediglich noch 7 l/m^3.

Jedoch muss berücksichtigt werden, dass beim Einsatz einer Betonpumpe wesentlich andere Pumpendrücke auf die Mischung einwirken als beim Versuch mit dem Sliper. Möglicherweise kann bei hohen Pumpendrücken in Betonpumpen die Luft komprimiert werden und somit nicht zur Ausbildung der Gleitschicht zur Verfügung stehen [13]. Untersuchungen von [14] zeigen einen äußerst positiven Einfluss auf die Pumpbarkeit durch die Einführung künstlicher Luftporen.

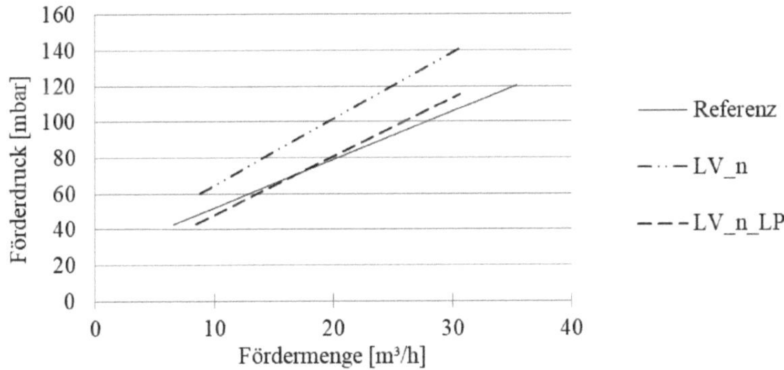

Bild 4: Pumpprognose ermittelt mit Sliper von Rezepturen mit unterschiedlichem Leimvolumen

4.2.2 Rheologische Eigenschaften des Leims

Die Einflüsse der rheologischen Eigenschaften des Leims auf die Pumpbarkeit können z. B. mit unterschiedlichen w/z-Werten dargestellt werden. In Tabelle 3 sind die Volumenanteile von Rezepturen mit unterschiedlichen w/z-Werten aufgeführt. Das Leimvolumen aus Zement, Wasser, Gesteinskörnung < 0,125 mm und Luft wurde annähernd konstant gehalten, um diesen Einflussfaktor möglichst auszuschließen.

Die Mischungen wurden mit einem Polycarboxylatether-Fließmittel auf ein Ausbreitmaß von 55 - 60 cm eingestellt (siehe Bild 5).

Tabelle 3: Volumenanteile von Rezepturen mit unterschiedlichen w/z-Werten

	Einheit	w/z=0,55	w/z=0,50 (Referenz)	w/z=0,45
CEM I 52,5 N SR0	[dm³/m³]	124	131	140
Wasser	[dm³/m³]	222	214	205
Gesteinskörnung < 0,125 mm	[dm³/m³]	38	38	38
Luftgehalt	[dm³/m³]	31	36	35
Gesamt Leimvolumen	[dm³/m³]	414	419	418

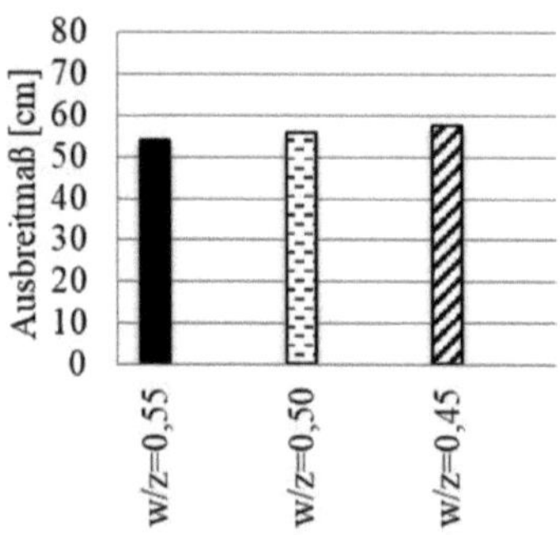

Bild 5: Ausbreitmaß von Rezepturen mit unterschiedlichen w/z-Werten

Die Ergebnisse der Sliperversuche zeigen deutlich, dass je geringer der w/z-Wert umso höher ist der zu erwartende Pumpendruck (siehe Bild 6) und umso höher ist die Viskosität der Mischungen bei annähernd gleichem Leimvolumen.

Die Mischungen mit unterschiedlichen w/z-Werten und damit unterschiedlichen Wassergehalten hatten ein ähnliches Ausbreitmaß, aber die Ergebnisse mit dem Gleitrohr-Rheometer ergaben unterschiedliche Pumpprognosen. Mit dem Gleitrohr-Rheometer konnte der Einfluss des Wassergehalts auf die Pumpbarkeit dargestellt werden.

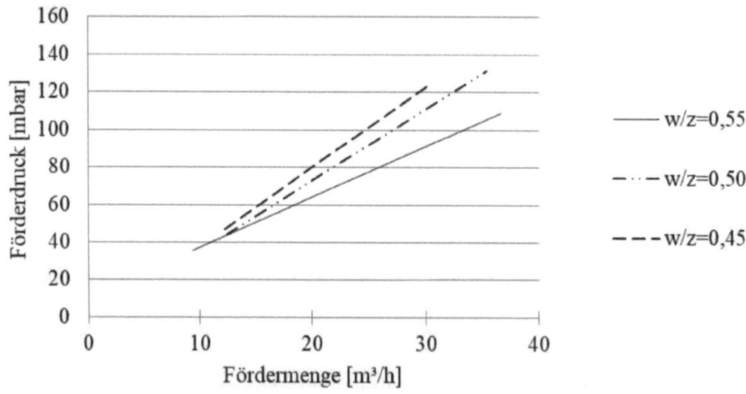

Bild 6: Pumpprognose ermittelt mit Sliper von Rezepturen mit unterschiedlichen w/z-Werten

Für die Beurteilung der Pumpbarkeit von Frischbetonmischungen ist neben den rheologischen Eigenschaften der Mischungen auch die Stabilität von Bedeutung. In dieser Versuchsreihe führte ein hoher w/z-Wert zu einer hohen Filtratwassermenge (siehe Bild 7). Eine geringe Viskosität der Mischung „w/z=0,55" ist günstig für niedrige Pumpendrücke, aber wenn die Mischung gleichzeitig zu einer hohen Wasserabgabe unter Druck neigt, besteht die Gefahr von Entmischungen und von Blockaden in den Förderleitungen.

Die Stabilität kann im Sliper messtechnisch nicht erfasst werden. Sie kann nur optisch durch die Ausbildung eines deutlich sichtbaren Wasserfilms an der Betonoberfläche und an der Rohrwandung erkannt werden oder wenn das Gleitrohr blockiert. Die Filterpresse ist ein geeignetes ergänzendes Verfahren, um die Mischungen differenzierter zu beurteilen.

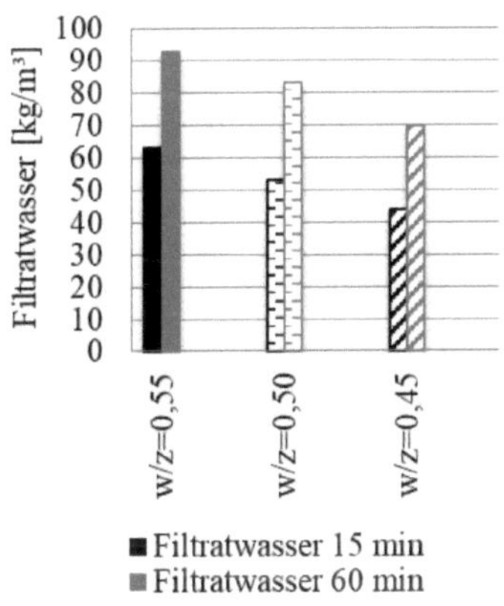

Bild 7: Filtratwasser nach 15 min und 60 min von Rezepturen mit unterschiedlichen w/z-Werten

5 Großspritzversuche

In Großspritzversuchen wurden die Erkenntnisse aus den Laborversuchen umgesetzt. Die Versuche fanden in einem Versuchsstollen statt und wurden mit einem Spritzmobil Sika - PM500 durchgeführt (siehe Bild 8).

Die Frischbetoneigenschaften der Grundmischung wurden vor jedem Spritzvorgang geprüft. Während dem Spritzen wurde der Hydraulikdruck der Zylinder aufgezeichnet, um Rückschlüsse auf die Pumpbarkeit zu erhalten.

Bild 8: Spritzmobil Sika - PM500 vor Versuchsstollen.

5.2 Referenzrezeptur

In Tabelle 4 ist die Rezeptur der Referenzmischung der Großspritzversuche aufgeführt. Es wurde ein im Rahmen des Forschungsprojektes „ASSpC" hinsichtlich Dauerhaftigkeit optimierter Zement „CEM II/B-M(S,L,Q) 52,5 N" eingesetzt.

Tabelle 4: Referenzrezeptur Großspritzversuche

	Dichte	Masse	Volumen
	[kg/dm³]	[kg/m³]	[dm³/m³]
CEM II/B-M(S,L,Q) 52,5 N	3,0	398	131
Wasser	1,0	197	197
Gesteinskörnung Dolomit 0/8	2,8	1811	638
Luftgehalt	---	---	32

Die Konsistenz aller nachfolgend beschriebenen Mischungen wurde mit einem Fließmittel auf ein Ausbreitmaß von 55 – 60 cm eingestellt. Mit den Rezepturvariationen wurde teilweise bewusst an die Grenzen der Verarbeitbarkeit gegangen, um Einflüsse auf die Pumpbarkeit herauszuarbeiten. Alle Spritzversuche wurden mit einer Fördermenge von 12 m³/h durchgeführt.

5.3 Ergebnisse Großspritzversuche

5.3.1 Zement- und Wasservolumen

In den Laborversuchen wurde festgestellt, dass das Leimvolumen einen Einfluss auf die Pumpbarkeit hat. Im Rahmen der Großspritzversuche wurde das Zement- und Wasservolumen im Vergleich zur Referenz systematisch reduziert. Das Leimvolumen aus Zement, Wasser, Gesteinskörnung < 0,125 mm und Luft ist bei allen Mischungen durch die Zugabe von Luftporenbildner ähnlich. Ziel dieser Versuchsreihe war es den Einfluss des Zement- und Wasservolumens auf die Ausbildung der Gleitschicht und damit auf die Pumpbarkeit zu untersuchen. In Tabelle 5 sind die Rezepturen der Referenzmischung und der Mischungen mit reduziertem Zement- und Wasservolumen enthalten.

Tabelle 5: Volumenanteile von Rezepturen mit reduziertem Zement- und Wasservolumen

	Einheit	Referenz	ZW_n_1	ZW_n_2
CEM II/B-M(S,L,Q) 52,5 N	[dm³/m³]	131	100	125
Wasser	[dm³/m³]	197	155	182
Gesteinskörnung < 0,125 mm	[dm³/m³]	46	46	46
Luftgehalt	[dm³/m³]	32	105	50
Gesamt Leimvolumen	[dm³/m³]	406	406	404

Bei den Mischungen „ZW_n_1" (**Z**ement- und **W**asservolumen_**n**iedrig_**1**) und „ZW_n_2" (**Z**ement- und **W**asservolumen_**n**iedrig_**2**) sind die Volumenanteile von Zement und Wasser deutlich geringer als bei der Referenzmischung. Der erhöhte Luftgehalt im Vergleich zur Referenzmischung wurde durch die Zugabe eines Luftporenbildners erreicht.

Die Pumpbarkeit der Mischungen wurde vor jedem Spritzvorgang mit dem Gleitrohr-Rheometer Sliper geprüft. In Bild 9 sind die Ergebnisse dargestellt.

Der zu erwartende Förderdruck von Mischungen mit einem geringen Zement- und Wasservolumen war deutlich höher als bei der Referenzmischung.

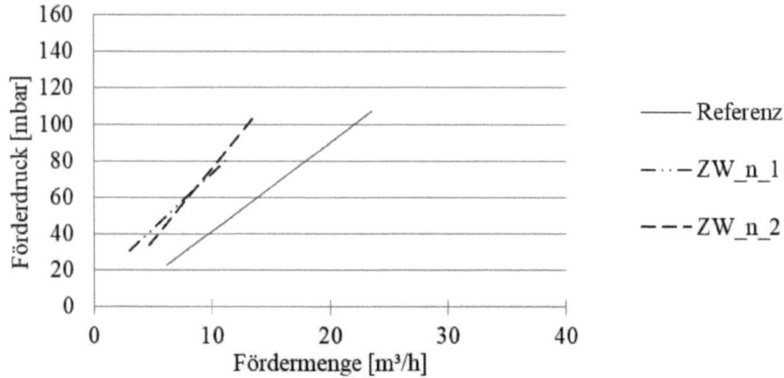

Bild 9: Pumpprognose ermittelt mit Sliper von Rezepturen mit reduziertem Zement- und Wasservolumen

Die Spritzversuche mit den Mischungen „ZW_n_1" und „ZW_n_2" mussten aufgrund der zu hohen Hydraulikdrücke abgebrochen werden, da ein Grenzwert von 200 bar zum automatischen Abbruch führte (siehe Tabelle 6). Der Hydraulikdruck war damit deutlich höher im Vergleich zu den Versuchen mit der Referenzmischung mit einem Hydraulikdruck von 70 bar. Das Volumen aus Zement und Wasser war bei den Mischungen „ZW_n_1" und „ZW_n_2" nicht ausreichend, um eine Gleitschicht auszubilden. Der hohe Luftgehalt konnte das fehlende Feststoffvolumen nicht ausgleichen. Die Ergebnisse belegen deutlich, dass trotz eines ähnlichen Ausbreitmaßes starke Unterschiede sowohl bei den Ergebnissen mit dem Gleitrohr-Rheometer als auch bei den Messungen des Hydraulikdrucks der Zylinder und damit der Pumpbarkeit vorhanden waren.

Ein Vergleich der mit dem Sliper ermittelten Viskosität (Beiwert b) und dem Hydraulikdruck der Zylinder zeigt, dass eine hohe Viskosität bei den Sliperversuchen zu einem hohen Hydraulikdruck geführt hat (siehe Tabelle 6).

Tabelle 6: Beiwert b (Viskosität ermittelt mit Sliper) von Rezepturen mit reduziertem Zement- und Wasservolumen, Hydraulikdruck der Zylinder und voreingestellte Fördermenge beim Spritzvorgang

	Beiwert b	Fördermenge	Hydraulikdruck Zylinder
	[10^{-6} bar*h/m]	[m³/h]	[bar]
Referenz	3,74	12	70
ZW_n_1	4,70	12	Abbruch > 200
ZW_n_2	6,02	12	Abbruch > 200

Bei der Beurteilung der Pumpbarkeit der Mischungen wurde zusätzlich die Filtratwassermenge mit einbezogen (siehe Bild 10).

Erwartungsgemäß führte das geringere Zement- und Wasservolumen der Mischungen „ZW_n_1“ und „ZW_n_2“ zu einer geringeren Filtratwassermenge.

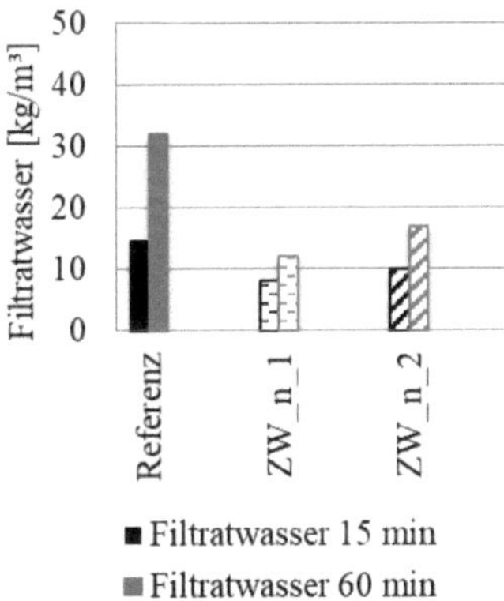

Bild 10: Filtratwasser nach 15 min und 60 min von Rezepturen mit reduziertem Zement- und Wasservolumen

5.3.2 Wassergehalt und Luftporengehalt

In weiteren Versuchen wurden der Wassergehalt und der Luftporengehalt im Vergleich zur Referenzrezeptur erhöht.

Tabelle 7: Volumenanteile von Rezepturen mit erhöhtem Wasser- und Luftporengehalt

	Einheit	Referenz	w/z_hoch	LP_hoch
CEM II/B-M(S,L,Q) 52,5 N	[dm³/m³]	131	132	133
Wasser	[dm³/m³]	197	208	196
Gesteinskörnung < 0,125 mm	[dm³/m³]	46	46	42
Luftgehalt	[dm³/m³]	32	15	90
Gesamt Leimvolumen	[dm³/m³]	406	401	461

Bei der Mischung „w/z_hoch“ beträgt der w/z-Wert 0,52, die Referenzrezeptur hat einen w/z-Wert von 0,49. Die Rezeptur „LP_hoch“ hat im Vergleich zur Referenz einen erhöhten Luftporengehalt durch die Zugabe von Luftporenbildner und damit ein höheres Leimvolumen.

In Bild 11 sind die mit dem Sliper ermittelten Pumpprognosen dargestellt.

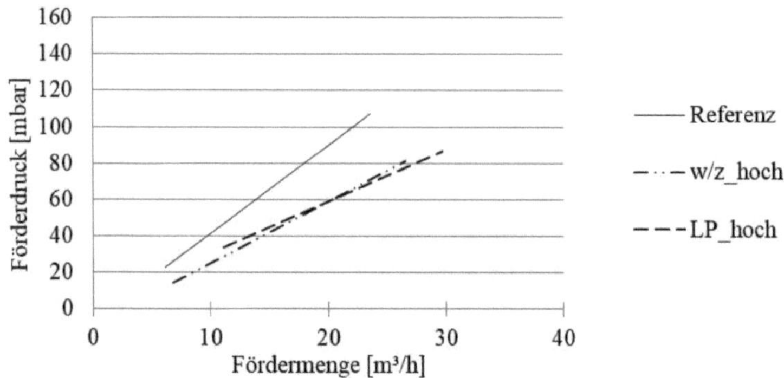

Bild 11: Pumpprognose ermittelt mit Sliper von Rezepturen mit erhöhtem Wasser- und Luftporengehalt

Sowohl der erhöhte Wassergehalt als auch der erhöhte Luftporengehalt führte im Vergleich zur Referenz zu einer Reduzierung des zu erwartenden Pumpendrucks.

Der während dem Spritzvorgang gemessene Hydraulikdruck der Zylinder war bei beiden Mischungen im Vergleich zur Referenz ebenfalls geringer und wirkte sich positiv auf die Pumpbarkeit aus (siehe Tabelle 8). Die mit dem Sliper gemessene Viskosität (Beiwert b) weist die gleiche Tendenz auf.

Die Messungen der Filtratwassermenge zeigen, dass bei dieser Versuchsreihe durch eine moderate Erhöhung des Wassergehalts oder eine Erhöhung des Luftporengehalts kein negativer Einfluss auf die Stabilität und damit die Pumpbarkeit zu erwarten war. Die Filtratwassermenge war bei allen drei Rezepturen ähnlich (siehe Bild 12).

Tabelle 8: Beiwert b (Viskosität ermittelt mit Sliper) von Rezepturen mit erhöhtem Wasser- und Luftporengehalt, Hydraulikdruck der Zylinder und voreingestellte Fördermenge beim Spritzvorgang

	Beiwert b	Fördermenge	Hydraulikdruck Zylinder
	[10^{-6} bar*h/m]	[m³/h]	[bar]
Referenz	3,74	12	70
w/z_hoch	2,60	12	63
LP_hoch	2,17	12	57

Die Beobachtungen beim Spritzvorgang und die optische Beurteilung der Mischungen hinsichtlich der Stabilität bestätigten diese Erkenntnisse.

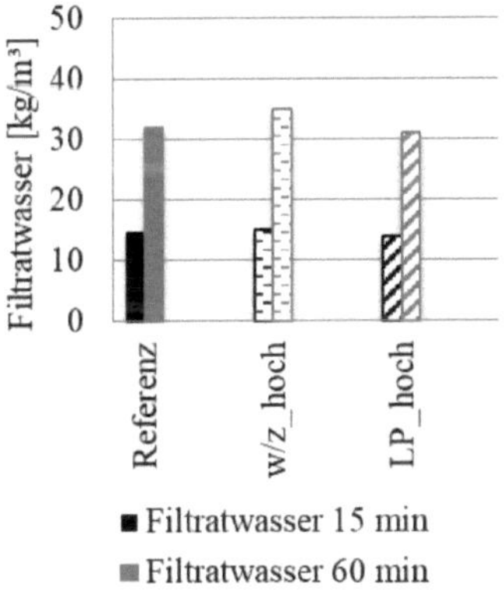

Bild 12: Filtratwasser nach 15 min und 60 min von Rezepturen mit erhöhtem Wasser- und Luftporengehalt

6 Zusammenfassung

In Labor- und Großspritzversuchen wurde die Verarbeitbarkeit von Frischbetonmischungen für das Nassspritzverfahren untersucht. Für die Prüfung der Mischungen im Labor und vor den Spritzversuchen wurden das Ausbreitmaß zur Beurteilung der Konsistenz, das Gleitrohr-Rheometer Sliper zur Beurteilung der Pumpbarkeit und die Filterpresse zur Beurteilung der Stabilität verwendet. Während der Spritzversuche wurde die Pumpbarkeit mithilfe des am Spritzmobil gemessenen Hydraulikdrucks der Zylinder beurteilt.

Folgende Erkenntnisse konnten aus den Labor- und Großspritzversuchen gewonnen werden:

- Das Ausbreitmaß war nicht geeignet um die Pumpbarkeit von Frischbetonmischungen zu beurteilen. Die mit dem Gleitrohr-Rheometer ermittelten Pumpprognosen hingegen ermöglichten in Kombination mit den Ergebnissen der Filterpresse zur Beurteilung der Stabilität Rückschüsse auf die Pumpbarkeit von Frischbetonmischungen.

- Das Leimvolumen, das Zement- und Wasservolumen sowie der Luftgehalt beeinflussten die Ausbildung der Gleitschicht und damit die Pumpbarkeit.

- Eine Erhöhung des Wassergehalts und/oder des Luftgehalts wirkte sich bei bereits pumpbaren Mischung positiv aus und führte zu geringeren Pumpendrücken, sofern die Mischungen stabil waren. Die Einführung künstlicher Luftporen konnte jedoch fehlendes Feststoffvolumen (Zement, Wasser und Feinanteile der Gesteinskörnung) zur Ausbildung der Gleitschicht nicht ersetzen.

7 Literatur

[1] Bartos, P.: Fresh Concrete - Properties and Tests. Amsterdam: Elsevier Science Publishers B.V. 1992.

[2] Kasten, K.: Gleitrohr-Rheometer. Ein Verfahren zur Bestimmung der Fließeigenschaften von Dickstoffen in Rohrleitungen. Dissertation. Dresden 2009.

[3] Kaplan, D., Larrad, F. de, Sedran, T.: Design of Concrete Pumping Circuit. ACI Materials Journal Volume 102 (2005), S. 110–117.

[4] DIN Deutsches Institut für Normung: DIN EN 14487-1: Spritzbeton - Teil 1: Begriffe, Festlegungen und Konformität. Berlin: Beuth Verlag GmbH (03.2006).

[5] Reinhold, M.: Methoden zur zielsicheren Vorhersage der Pumpbarkeit von Betonen mit nicht idealer Gesteinskörnungs-Sieblinie. BetonWerk International 1 (2014), S. 56–59.

[6] Secrieru, E., Butler, M., Mechtcherine, V.: Prüfen der Pumpbarkeit von Beton - Vom Labor in die Praxis. Bautechnik 91 (2014), S. 797–811.

[7] Ngo, T. T., Kadri, E. H., Cussigh, F., Bennacer, R.: Relationship between concrete composition and boundary layer composition to optimise concrete pumpability. European Journal of Environmental and Civil Engineering 16 (2012), S. 157–177.

[8] Jacobsen, S., Hâvard, M., Foon Lee, S., Haugan, L.: Pumping of concrete and mortar - State of the art. COIN Project report 5. Oslo 2008.

[9] DIN Deutsches Institut für Normung: DIN EN 12350-5: Prüfung von Frischbeton - Teil 5: Ausbreitmaß. Berlin: Beuth Verlag GmbH (08.2009).

[10] Schleibinger: Sliper.

http://www.schleibinger.com/cmsimple/en/?Rheometers%26nbsp%3B:.recent_news:SLIPER_%96_Determining_the_pumping_capacity_of_concrete.

[11] Österreichische Vereinigung für Beton- und Bautechnik: Merkblatt Weiche Betone. Wien (12.2009).

[12] Österreichische Vereinigung für Beton- und Bautechnik: Richtlinie Spritzbeton. Wien (12.2009).

[13] Jolin, M., Burns, D., Bissonette, B., Gagnon, F., Bolduc, L.-S.: Understanding the Pumpability of Concrete. In: ECI Conference (Ed.): Proceedings Shotcrete for Underground Support XI. Davos, 2009.

[14] Beaupré, D.: Rheology of High Performance Shotcrete. PhD Thesis. Vancouver 1994.

Mischer für klassische Frischbetonversuche

Labor – Intensivmischer

KKM – RT 15 / 22,5

\usbreitmaß
;etzfließmaß
'dZ-Trichter
.CPC-Box

dem integrierten **Rheometer** und **Tribometer** öglicht der KKM-RT die Eigenschaften des :hguts in relativen und absoluten Einheiten zu sen.

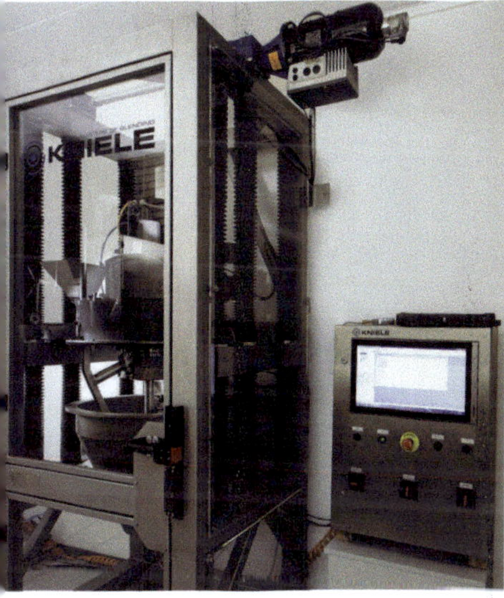

Features:

Rheometer und Tribometer

- Messung absolut und relativ

Spezialrührwerk

- Aufschluss von Desagglomeraten

Mischersonde

- Messung der Feuchte

Kamera

- Beobachtung des Mischvorgangs

eter mit Schnellwechselsystem

Spezialrührwerk

Mischersonde

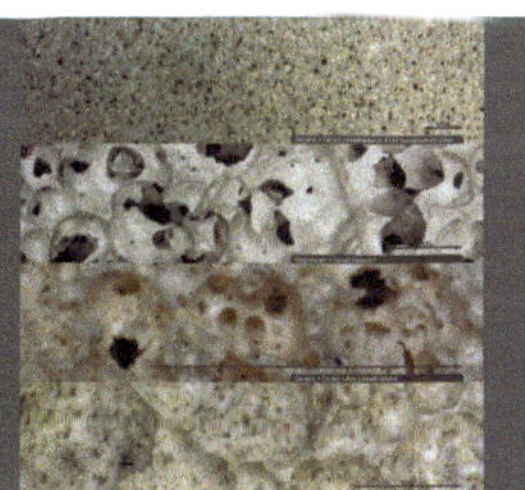

Laborversuche mit Schaumbetonen

le GmbH
eindebeunden 6
22 Bad Buchau
: +49 7582 9303 - 0
: +49 7582 9303 - 30
: info@kniele.de

blending • dosing • conveying – made in Germany

kniele.de